9th International Conference on Mechatronics and Manufacturing (ICMM 2018)

AA006904

IOP Conference Series: Materials Science and Engineering
Volume 361

Phuket, Thailand
27 – 29 January 2018

ISBN: 978-1-5108-6399-6
ISSN: 1757-8981

Printed from e-media with permission by:

Curran Associates, Inc.
57 Morehouse Lane
Red Hook, NY 12571

Some format issues inherent in the e-media version may also appear in this print version.

Copyright© (2018) by the Institute of Physics
All rights reserved. The material featured in this book is subject to
IOP copyright protection, unless otherwise indicated.

Printed by Curran Associates, Inc. (2018)

For permission requests, please contact the Institute of Physics
at the address below.

Institute of Physics
Dirac House, Temple Back
Bristol BS1 6BE UK

Phone: 44 1 17 929 7481
Fax: 44 1 17 920 0979

techtracking@iop.org

Additional copies of this publication are available from:

Curran Associates, Inc.
57 Morehouse Lane
Red Hook, NY 12571 USA
Phone: 845-758-0400
Fax: 845-758-2633
Email: curran@proceedings.com
Web: www.proceedings.com

Table of contents

Volume 361

9th International Conference on Mechatronics and Manufacturing (ICMM 2018)

27–29 January 2018, Phuket, Thailand

Accepted papers received: 2 May 2018
Published online: 4 June 2018

Preface

Papers

Engineering Materials Modeling, Analysis and Testing

Mechanical System Design and Simulation Calculation

Mechatronics

Industrial Control and Automation

IOP Conf. Series: Materials Science and Engineering **361** (2018) 011001 doi:10.1088/1757-899X/361/1/011001

Preface

2018 9th International Conference on Mechatronics and Manufacturing (ICMM 2018) took place Phuket, Thailand during Jan 27-29, 2018. ICMM 2018 aims to bring together researchers, scientists, developers, and engineers from around the world in both industry and academia for sharing state-of-art research results, for exploring new areas of research and development, and to discuss emerging issues in the fields of mechatronics and manufacturing. It provides opportunities for the delegates to exchange new thoughts and application experiences, to establish business or research relations and to find global partners for possible future collaboration, and also it helps expanding the delegates' knowledge and insight into the significant challenges currently being addressed in the related fields.

The proceedings of ICMM 2018 records the fully refereed papers presented at the conference, which include 26 excellent papers selected from more than 51 submitted papers whose authors are from over 9 countries. All papers published in the proceeding have been peer reviewed by the reviewers drawn from the technical committee members and external reviewers depending on the subject matter of the paper. Its scope includes but not limited to: Engineering Materials Modeling, Analysis and Testing, Mechanical System Design and Simulation Calculation, Mechatronics and Industrial Control and Automation. After the rigorous peer-review process, the submitted papers were selected on the basis of originality, significance, and clarity for the purpose of the conference.

We would like to thank the program chairs, reviewers and the members of the program committee for their hard work. We are grateful to all those who have contributed to the success of ICMM 2018. We hope that all participants and other interested readers benefit scientifically from the proceedings.

ICMM 2018 Organizing Committees

February 12, 2018

Content from this work may be used under the terms of the Creative Commons Attribution 3.0 licence. Any further distribution of this work must maintain attribution to the author(s) and the title of the work, journal citation and DOI.
Published under licence by IOP Publishing Ltd

IOP Conf. Series: Materials Science and Engineering **361** (2018) 011002 doi:10.1088/1757-899X/361/1/011002

Conference Committee

Conference Chair

Prof. Yoshihiko Uematsu,Gifu University, Japan

Prof. Po-Liang Liu, National Chung-Hsing University, Taiwan

Program Chair

Prof. Vadim R. Gasiyarov, South Ural State University, Russia

Technical Committee

Prof. Dong-Won Jung, Jeju national University, Republic of Korea

Assoc.Prof. Tapas Kumar Maiti, Hiroshima University, Japan

Prof. Osman Adiguzel, Firat University, Turkey

Prof. Yusri Yusof, Universiti Tun Hussein Onn Malaysia (UTHM) , Malaysia

Prof. Hamid Reza Karimi, Polytechnic University of Milan, Italy

Prof. Shi Yuliang, Beijing University of Technology, China

Prof. Prakash.S, Sathyabama University, India

Prof. N.ETHIRAJ, Dr. M.G.R Educational and Research Institute, Chennai, India

Prof. A A Sandulyak, Moscow Technological University, Russia

Prof. Minsoo Park, Seoul National University of Science and Technology, South Korea

Assoc.Prof. K Kanlayasiri, King Mongkut's Institute of Technology Ladkrabang, Thailand

Assoc.Prof. Shih-Chuan Yeh, De Lin Institute of Technology, Taiwan

Assoc.Prof.Shailendra Kumar, S.V.National Institute of Technology, India

Assoc.Prof. Chia-Hung Wei, UCH, Taiwan

Assoc. Prof. Dondee Navarro, Hiroshima University, Japan

Assoc.Prof. Nazar Elfadil Mohamed, Fahad bin Sultan University, Kingdom of Saudi Arabia

Dr. K.K.Soundrapandian, Pdpm-Indian Institute Of Information Technology Design And Manufacturing, India

Asst.Prof. Elmer R. Magsino, De La Salle University, Philippines

Asst.Prof. Wandee Petchmaneelumka, King Mongkut's Institute of Technology Ladkrabang, Thailand

Asst.Prof. Pek-Lan Toh, Universiti Tunku Abdul Rahman, Malaysia

Dr. Yang Xi, Thinfilim Electronics Inc., U.S.A

Content from this work may be used under the terms of the Creative Commons Attribution 3.0 licence. Any further distribution of this work must maintain attribution to the author(s) and the title of the work, journal citation and DOI.
Published under licence by IOP Publishing Ltd

IOP Conf. Series: Materials Science and Engineering **361** (2018) 011003 doi:10.1088/1757-899X/361/1/011003

Peer review statement

All papers published in this volume of *IOP Conference Series: Materials Science and Engineering* have been peer reviewed through processes administered by the proceedings Editors. Reviews were conducted by expert referees to the professional and scientific standards expected of a proceedings journal published by IOP Publishing.

Content from this work may be used under the terms of the Creative Commons Attribution 3.0 licence. Any further distribution of this work must maintain attribution to the author(s) and the title of the work, journal citation and DOI.
Published under licence by IOP Publishing Ltd

IOP Conf. Series: Materials Science and Engineering **361** (2018) 012001 doi:10.1088/1757-899X/361/1/012001

Influences of solid-state brazing conditions on microstructure and tensile shear force of Ag-Cu-P brazed copper sheets

P Jattakul and K Kanlayasiri

Department of Industrial Engineering, Faculty of Engineering,
King Mongkut's Institute of Technology Ladkrabang, Bangkok 10520, Thailand

Email: Prajak_Jattakul@yahoo.com

Abstract. This study aimed to examine the tensile shear force and microstructure of copper sheets when Ag-Cu-P filler metal was applied for the joints involved in the production of laminated micro-channel array (LMA) devices. The properties of the microstructure of the Ag-Cu-P/copper brazed joints were examined following a process which included furnace brazing, with the parameters comprising temperature, holding time, and loading pressure. Analytical tools included an energy dispersive spectrometer (EDS), scanning electron microscope (SEM) and X-ray diffraction analysis (XRD). In addition, tensile tests were performed to measure the tensile shear force. It was difficult to observe the Cu_3P which forms along with the layer between the Ag-Cu-P filler metal and the copper sheet when the brazing temperature was only 580°C since this allowed gaps to form within the joint, thus producing very low tensile shear force results. When the brazing temperature is increased to 620°C along with longer holding time and increased pressure, this allows a thicker phase of Cu_3P to form in the brazing joint. With a holding time of 30 minutes and pressure of 12.173 kPa, tensile shear force was maximized since this condition produced thick Cu_3P phase. It is apparent that it is possible to produce a brazed joint which has an average shear force of 736.27 N under these conditions as described above. Within the joint could be found a Cu_3P phase with no brittle phase.

1. Introduction

The large-scale processing of mass and energy through laminated micro-channel array (LMA) devices is known as micro-channel process technology (MPT). [1–2]. In comparison with conventional fluidic technology the most significant advantage of MPT is the very high the ratio of surface area to volume, permitting heat and mass transfer to take place very quickly within the micro-channels as the distances over which diffusion taking place are much shorter. This allows micro-channels to provide energy and chemical systems of reduced weight and size [3]. MPT is applied to an ever greater extent in numerous industrial application, such as the recuperation of waste heat, and portable heat exchanger [4].

In recent years, MPT technology has been made using mechanical processes, or by selective laser melting. Under mechanical methods, the structures are made by cutting sheets very accurately and then joining the stacked sheet layers to make the MPT device by fusion joining or by brazing and diffusion bonding [5]. In fusion welding, sheets are joined around the edges, ensuring that the micro-channels are not inevitably in direct contact with each other, and thus the total welding area is not large. These particular MPT devices are not expected to endure high pressure, while a further weakness is the development of the heat affected zone (HAZ), which can undergo considerable change to the bonded material microstructure, which ultimately causes the loss of mechanical properties [6].

Content from this work may be used under the terms of the Creative Commons Attribution 3.0 licence. Any further distribution of this work must maintain attribution to the author(s) and the title of the work, journal citation and DOI.
Published under licence by IOP Publishing Ltd

IOP Conf. Series: Materials Science and Engineering **361** (2018) 012001 doi:10.1088/1757-899X/361/1/012001

Earlier studies have presented a number of methods such as, diffusion bonding or fusion welding. These approaches have all been used to form the joints in manufacturing MPT devices [7–8]. The most widely used approach in MPT production is diffusion bonding. This method is widely used because it ensures that the micro-channels are tightly held together. In addition, there is no liquid phase involved in the process, and the procedure can employ parameters which do not result in changes to the microstructure. However, lengthy bonding time is required for diffusion bonding.

In this study, the solid-state diffusion brazing of copper sheets using Ag-Cu-P filler metal is proposed as another alternative for LMA fabrication. The research involves brazing using varying parameters of holding time, brazing temperature, and loading pressure. The microstructure and tensile shear force of the brazed joint were assessed using SEM and EDS, while tensile tests were performed to evaluate the tensile shear force of the joint. The effects of changes in the three parameters upon the characteristics of the joints were subjected to detailed examination. The relationship between the microstructure and the mechanical properties of the joint is discussed in this paper. The results of this study should prove to be useful in establishing the set of manufacturing processes which can subsequently be employed in the development of a range of LMA devices.

2. Experiment

2.1. Materials and methods

For the solid-state brazing, the base material used in the study was commercially pure copper at a thickness of 0.350 mm. The samples were divided into rectangular plates with 10×27.7×0.35 mm in dimensions. Ag-Cu-P with a thickness of 0.250 mm was employed as the filler metal. In the experiment, rectangular plates of the filler metal were cut to a size of 10×5×0.25 mm. Furnace brazing was carried out using a BOREL Swiss SA No. S0908 model. The solid-state brazing process was performed under a hydrogen atmosphere to prevent the oxidation. Following the solid-state brazing process, the microstructure was examined and the composition of the brazing joints was assessed using SEM (JEOL model JSM-6510LV, Japan). An EDS (Inca, Oxford, UK) was also employed for the analysis. Testing of the tensile shear force of the brazing joints was carried out at room temperature and a crosshead speed of 0.1 mm/min, using Shimadzu AG-100 universal testing machine. The phases in the filler metal and brazed joints were identified using XRD (Bruker D8 Discover), and Cu Kα was chosen as the X-ray source.

3. Results and Discussion

3.1. Ag-Cu-P filler metal microstructure

The microstructure of Ag-Cu-P filler metal is presented in Figure 1 (a-e). Among the phases shown, the white one is associated with silver, the light gray color comprises of copper, and the dark gray area is Cu_3P compound. Chemical composition analysis can be employed to determine the exact constituents and distribution of the major phases, as displayed in the Table 1.

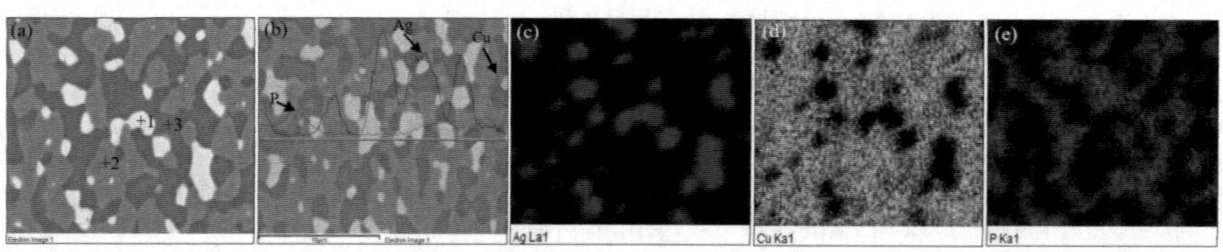

Figure 1. (a) Characteristics of the Ag-Cu-P filler metal microstructure; (b) Profile analyses of the filler metal by EDS linear scanning; Element distribution maps for the filler metal generated via EDS, including (c) Silver; (d) Copper and (e) Phosphorus.

3.2. Brazed joint microstructure characterization

The microstructure of the brazed joint between the copper sheet and the filler metal is presented in Figure 2. The brazing conditions consisted of a temperature of 620°C for 30 minutes at a loading pressure of 12.173 kPa. In this case the wetting of the filler metal with the copper sheet was very good, and no holes or cracks were found in the joint as shown Figure 2 (a-b). There are two distinct parts of the brazing joint, which are denoted using the Roman numerals I and II as shown in Figure 2 (a-b). The section marked as I consists of the continuous reaction layer which is nearest to the copper sheet, while the section marked II forms the center of the joint. The two parts, I and II consist mainly of the silver phase (in white, denoted by 1), while the copper phase is light gray denoted by 2, and the dark gray phosphorus phase is shown at point 3. The compositions of the marked zones shown in Figure 2 (b) tabulated in Table 1.

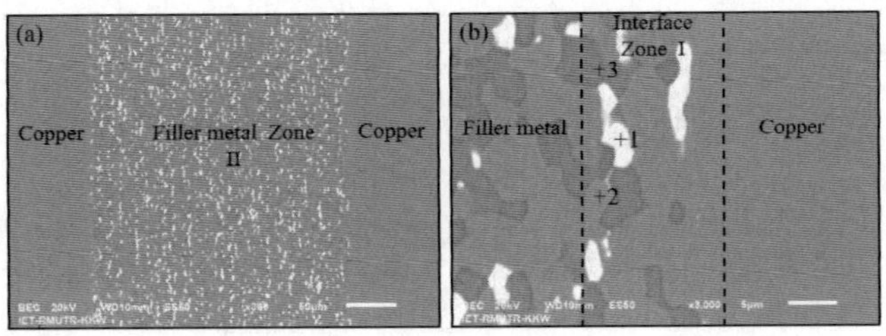

Figure 2. Brazed joint morphology (a-b), including (b) marked zones of the chemical composition for a brazed joint under conditions of 620°C, 30 minutes, and 12.173 kPa.

Table 1. The compositions of the marked zones from Figure 2 (b) and Figure 4 (a, f and k).

Solid-state brazing condition	Position	Composition (at. %)			Possible phase
		Ag	Cu	P	
Filler metal	1	89.00	10.44	0.56	Ag
	2	0.92	96.03	3.05	Cu
	3	0.05	70.36	29.59	Cu_3P
580°C; 10 min; 8.656 kPa	1	64.09	28.88	7.03	Ag
	2	1.47	95.87	2.66	Cu
	3	8.27	70.12	21.61	P
600°C; 30 min; 12.173 kPa	1	56.37	42.97	0.65	Ag
	2	0.79	79.70	19.51	Cu
	3	1.66	85.79	12.54	P
620°C; 30 min; 12.173 kPa	1	85.35	14.07	0.58	Ag-rich
	2	1.85	95.35	2.80	Cu-rich
	3	0.57	74.43	25.01	Cu_3P

Figure 3 (a-d) presents the EDS results in the form of compositional maps of the brazed joint along with the linear scanning profiles. The elemental spatial distribution can thus be specified. It is apparent that the brazed joint contains a continuous Ag-rich phase, a phase which is Cu-rich and the Cu_3P phase. On the basis of the binary Cu-P phase diagram along with the atomic ratio of the microstructure of the brazed joint, it can be stated that the reaction layer may comprise both Cu_3P and CuP_2 phase [9], However in this study there was no evidence of the formation of CuP_2 phase at the brazed joint. In this case it can be seen from the binary phase diagram for Ag-Cu that there was no intermetallic phase, thus confirming that Ag-Cu formed a solid solution and indicating that brazing can take place without the formation of an intermetallic phase. In the formation of Cu_3P, it is was occurred from phosphorus atoms in filler metal diffused towards the interface, and copper atoms in base material diffused towards the interface, thus leading to the formation of Cu_3P compound.

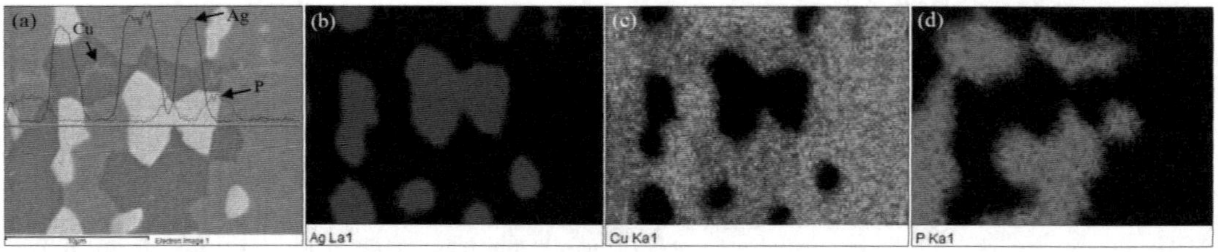

Figure 3. Distribution images of the corresponding elements area for brazed copper joints using Ag-Cu-P filler metal at 620°C for 30 min and 12.173 kPa: (a) Linear scanning profiles; (b) Ag phase distribution map; (c) Cu phase, and (d) Cu_3P compound.

3.3. The influence of parameters for brazing temperature, holding time and loading pressure upon joint microstructure

Figure 4 (a, f and k) presents the SEM micrographs of the brazed copper joints which used Ag-Cu-P filler metal. The temperature, holding time, and loading pressure varied in each case. It is also apparent that the joint between the Cu substrate and the filler metal contained micro-voids throughout the interface, calling into question its efficiency, as shown in Figure 4 (a). This occurred as a result of the conditions: at a low temperature, short holding time, and low loading pressure (580°C for 10 min and 8.656 kPa), it is not possible for the filler alloy to fully react with the copper sheet to create the continuous reaction layer. When the solid-state brazing conditions are improved (600°C for 30 min and 12.173 kPa) the voids become less significant and it is possible to distinguish a continuous reaction layer, as shown in Figure 4 (f). When the conditions are further intensified (620°C for 30 min and 12.173 kPa) the cracks disappear altogether, as shown in Figure 4 (k). Furthermore, the grain size of the region phase showed clear growth, and thus it can be inferred that the filler metal was able to fully react with the copper to create the continuous reaction layer, and consequently to form a Cu_3P compound (dark gray) of a suitable thickness.

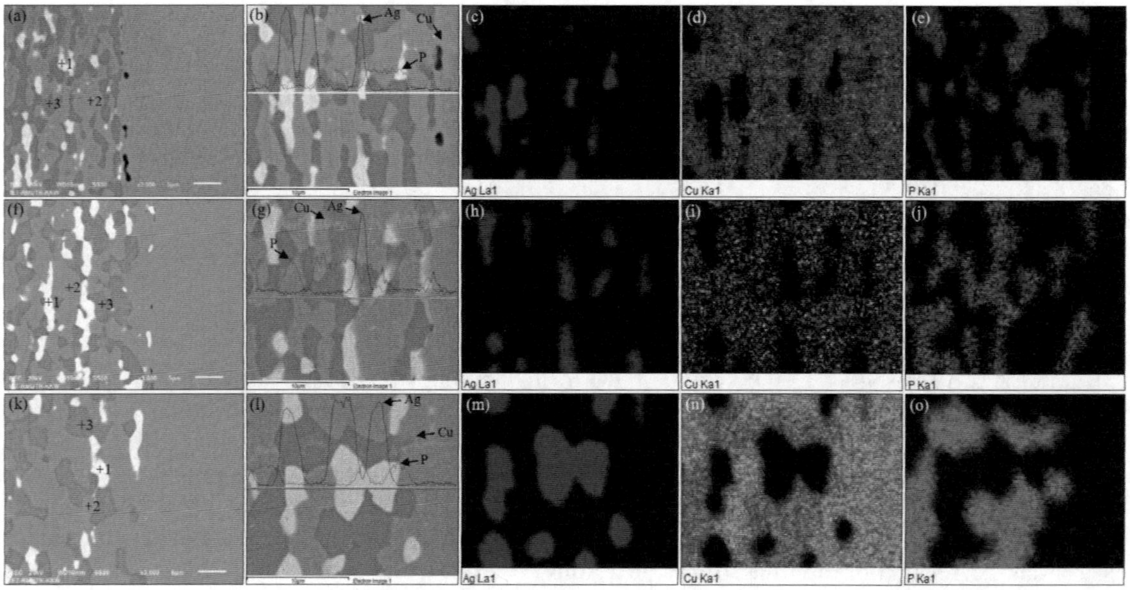

Figure 4. SEM micrographs showing the linear scanning profile and profiles of the EDS compositional maps of the brazed copper joint using Ag-Cu-P filler metal under varying solid-state brazing conditions: (a-e) 580°C for 10 min and 8.656 kPa; (f-j) 600°C for 30 min and 12.173 kPa, and (k-o) 620°C for 30 min and 12.173 kPa.

Figure 4 (b-e, g-j and l-o) presents maps showing EDS composition and linear scanning profiles for brazed joints formed under varying solid-state brazing conditions. The images show the formation of Cu-rich and Ag-rich phases along with Cu_3P upon the brazed joints. The details of the elemental composition of the results from different brazing conditions are provided in Table 1 and can be seen also in Figure 4 (l-o). The result confirms that the alloy primarily consists of Cu-based solid solution, Ag-based solid solution and Cu_3P compound. Figure 5 (a-d) shows the XRD pattern of filler metal and brazing specimens after solid-state brazing with copper sheet on the experimental brazing variables encompassed the brazing temperature, holding time and loading pressure at 580°C for 10 min and 8.656 kPa, 600°C for 30 min and 12.173 kPa and 620°C for 30 min and 12.173 kPa respectively. Obviously, Cu_3P is formed from the beginning of forming process of the filler metal. However, after solid-state brazing at high temperature (620°C for 30 min and 12.173 kPa) grains of Cu_3P phase were larger.

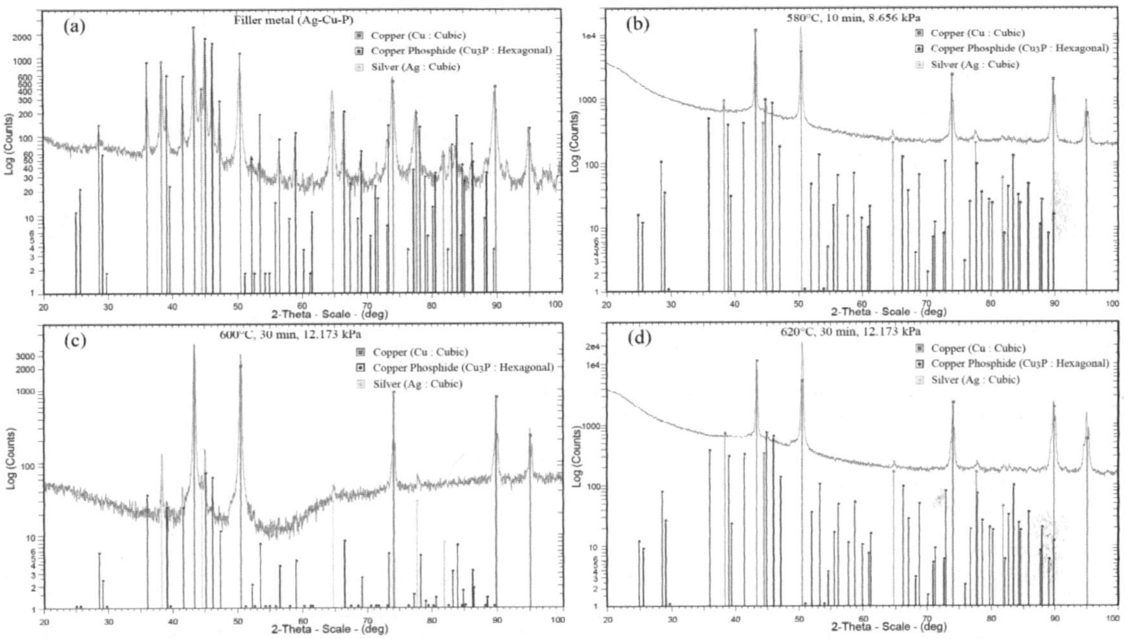

Figure 5. XRD pattern of filler metal and brazed joints after solid-state diffusion brazing under variable of brazing temperature, holding time and loading pressure respectively. (a) XRD pattern of filler metal, (b) XRD pattern of brazed joint after solid-state brazing at 580°C for 10 min and 8.656 kPa, (c) 600°C for 30 min and 12.173 kPa and (d) 620°C for 30 min and 12.173 kPa.

3.4. The mechanical characteristics of the joints

Evaluation of the tensile shear force was carried out in order to determine the properties of the brazed joint. The various brazing conditions involved temperatures of 580°C, 600°C, and 620 °C; 10 min, 20 min, and 30 min, and 4.713 kPa, 8.656 kPa and 12.173 kPa. There is a clearly apparent increase in the tensile shear force as the temperature, holding time, and loading pressure are increased. Analysis of the microstructure revealed that voids were present in the joints at low temperatures, short holding times, and low loading pressure, indicating relatively weak bonding (Figure 4 (a and f)). At higher temperatures and loading pressures with a longer holding time, the voids disappear and stronger joints are the result.

In the confirmation test process, a total of six samples were brazed. Table 2 shows the outcome in the high brazing sample, representing the optimal brazing conditions. The tests confirmed that all samples brazed under optimal conditions achieved tensile shear force results which are considered to fall within the range of higher tensile shear force.

IOP Conf. Series: Materials Science and Engineering **361** (2018) 012001 doi:10.1088/1757-899X/361/1/012001

Table 2. Testing to confirm the optimal solid-state brazing conditions.

Specimen (no.)	Tensile shear force (N)	Specimen (no.)	Tensile shear force (N)
1	743.771	4	744.565
2	746.838	5	713.015
3	731.611	6	737.842

4. Conclusions

The current study assessed the properties and microstructure of the brazed joints between copper sheets and Ag-Cu-P filler metal which are potentially employed in the production of LMA devices. The process involved solid-state brazing at temperatures of 580°C, 600°C and 620°C, with loading pressure at 4.713 kPa, 8.656 kPa, and 12.173 kPa, and holding times of 10 min, 20 min, and 30 min. The findings allowed certain conclusions to be proposed, as described below:

To obtain the greatest tensile shear force on average, the parameters were set as follows: brazing temperature 620°C; holding time 30 minutes; loading pressure 12.173 kPa. This resulted in average tensile shear force of 736.27 N. The analysis using EDS and SEM showed that it was difficult to form a reaction layer between the copper sheet and the Ag-Cu-P filler metal at the lower temperatures of 580°C and 600°C, lower holding time of 10 min and 20 min and lower loading pressure of 4.713 kPa and 8.656 kPa leading to gaps in the joint and low shear force. However, under optimal conditions (brazing temperature 620°C; holding time 30 minutes; loading pressure 12.173 kPa) Cu_3P phase was produced and apparent at the brazing joint between the filler metal and the copper sheets, leading to much better results on tensile shear force and a stronger joint.

References

[1] Paulraj P and Paul B K 2011 Metal microchannel lamination using surface mount adhesives for low-temperature heat exchangers *Journal of Manufacturing Processes* vol 13 pp 85–95.

[2] Paul B K and Lingam G K 2012 Cooling rate limitations in the diffusion bonding of microchannel arrays *Journal of Manufacturing Processes* vol 14 pp119–125.

[3] Paul B K, Kwon P and Subramanian R 2006 Understanding limits on fin aspect ratios in counter flow microchannel arrays produced by diffusion bonding. *Journal of Manufacturing Science and Engineering* vol 128 pp 977–983.

[4] Eluri R and Paul B K 2012 Silver nanoparticle-assisted diffusion brazing of 3003 Al alloy for microchannel applications. *Materials and Design* vol 36 pp 13–23.

[5] Kanlayasiri K, Paul B K 2004 A nickel aluminide microchannel array heat exchanger for high-temperature applications *Journal Manufacturing Process* vol 6 pp 17–25.

[6] Basuki W W, Kraft O and Aktaa J 2012 Optimization of solid-state diffusion bonding of Hastelloy C-22 for micro heat exchanger applications by coupling of experiments and Simulations *Materials Science and Engineering* A vol 538 pp 340–348.

[7] Eluri R and Paul B K 2012 Silver nanoparticle-assisted diffusion brazing 3003 Al alloy for microchannel applications *Materials and Design* vol 36 pp 13–23.

[8] Cooke K O, Khan T I and Oliver G D 2012 Transient liquid phase diffusion bonding Al-6061 using nano-dispersed Ni coatings *Materials and Design* vol 33 pp 469–475.

[9] Okamoto H, Subramanian P R, Kacpzak L and Massalski T B 2001 Binary alloy phase diagram of Cu-P Second edition *The Materials Information Society* vol 2 pp 1449–1451.

Acknowledgements

The researchers are grateful to Ronda (Thailand) Co. Ltd. who provided the brazing fixtures and the access to the brazing furnace, and would also like to extend sincere thanks to the Faculty of Engineering at King Mongkut's Institute of Technology Ladkrabang for providing financial assistance to this research.

IOP Conf. Series: Materials Science and Engineering **361** (2018) 012002 doi:10.1088/1757-899X/361/1/012002

The Comparison of the AST 908, AST 915 and AST 928 of Micro-Hardness of Tungsten Based Alloy Films on AISI 1050 Steel by HVOF Coating Method

P. Kumkoon, P. Surin

Faculty of Engineering, Pathumwan Institute of Technology, Bangkok, 10300, Thailand

Email : Piyapong_watpt@hotmail.com

Abstract. In this study, the commercial powder tungsten based alloys was sprayed by using high velocity oxygen fuel (HVOF) method on AISI 1050 steel substrate. These films of powder tungsten based alloys of AST 908, AST 915 and AST 928 were included. The powder was mainly composed of tungsten (W), Chromium (Cr), Nickel (Ni) and Cobalt (Co) with an average particle size in the range of 10-45 μm. The microstructure and compositional analysis was used by the scanning electrode microscopy (SEM) and energy dispersive spectroscopy (EDS) respectively. The micro-hardness of coating films was investigated by Vickers micro-hardness tester. The SEM results exhibited good deposition between the coating films with the specimens (AISI 1050 steel). Accordingly, the coating films showed a very homogeneous microstructure. In addition, the micro-hardness of substrate was 81.30% lower than the AST 908 particles. Finally, the AST 908, AST 915 and AST 928 coating improved the micro-hardness of 1559.3 HV, 1468.4 HV and 1306.2 HV respectively

1. Introduction

Industry was the activities of processing or manufacturing procedure of raw materials to make new materials to be useful according to the purpose of using machines or labour for mass production which could be sold as a product. A main factor of the production of each factory was the machinery for production. When these machines had been used for the mass production for a long time, [1] there were some parts of the machines that would be damaged or worn out after using them. Industrial examples in most countries such as steel smelting industry, automotive industry and others had large machines. When these parts of the machines were damaged that they couldn't avoid for machinery and engineering structures, selecting suitable materials for using and processing improvement a quality of the material was utilized to increase efficiency. As well as an understanding of the nature of wornness and eligible maintenance, it was a necessity for those accomplices of usage and maintenance for the wornness of some parts of industrial machines which could be repaired by various methods and brought to come back to life. The damaged parts were welded to the maintenance. It was an interesting way to use the heat-treated coating process which could be applied to many types of work. Spray coating with thermal flame was a useful technique in various industries. [2-3]. There are electronics, automotive, turbine and aircraft industries. This technique was often used in preventing wornness, heat and corrosion in chemicals and then used for repairing worn-out pieces. We could choose an infinite coated materials to get the desired properties. The used materials for the coating were metals, alloys,

Content from this work may be used under the terms of the Creative Commons Attribution 3.0 licence. Any further distribution of this work must maintain attribution to the author(s) and the title of the work, journal citation and DOI.
Published under licence by IOP Publishing Ltd

ceramics, composites and polymers. Spray coating with heat flame had a simple principle by heating with materials for coating on the surface that could be a powder or wires [4]. The thermal spraying technology was a coating method which was used extensively for the application on manufacturing and maintenance in Thailand. This research was studied the thermal spraying technology of worn-out resistance for tool life expansion.The aim of the study is to characterize the microstructure of tungsten based alloy films on AISI 1050 steel with different three-type powders namely; AST 908, AST 915 and AST 928 respectively.

2. Experimental procedure

In this study, high velocity oxygen fuel (HVOF) spraying method was coated on the specimen (AISI 1050 steel). The dimension of the specimen was 25.4 x 50.0 x 6.0 mm and chemical composition of specimen as shown in Table 1. The commercial powder tungsten based alloy namely AST 908, AST 915 and AST 928 was selected as coating materials. The powder tungsten based alloy was mainly composed of tungsten (W), chromium (Cr), nickel (Ni) and cobalt (Co) with an average particle size in the range of 10-45 μm. These commercial powder tungsten based alloy of experiment were AST 908, AST 915 and AST 928 which was shown in Table 2.

Table 1. The chemical composition and hardness value of substrate material.

Specimen	Chemical composition (%wt)					Hardness (HV)
	C	Mn	P	S	Si	
AISI 1050	0.47-0.55	0.60-0.90	0.020-0.035	0.020-0.035	0.15-0.35	247

Table 2. The chemical composition of commercial powder tungsten based alloy.

Type of powder	Chemical composition (%wt)			
	W	Co	Cr	Ni
AST 908	73	-	20	7
AST 915	88	12	-	-
AST 928	80	6	10	4

The parameters of HVOF coating optimized process were illustrated according to Table 3. Before HVOF coating process, the prepared specimens of surface properties were polished with silica-paper and cleaned with acetone. This objective of study was to compare the micro-hardness of tungsten based alloy films on AISI 1050 steel by HVOF coating method. Three films of powder tungsten based alloys were AST 908, AST 915 and AST 928 respectively.

Table 3. HVOF coating process parameters.

Parameters	Type of tungsten based alloy		
	AST 908	AST 915	AST 928
Oxygen pressure (ft^3/min)	1900	2000	1950
Fuel (gal/min)	6	6	5.75
Nitrogen pressure (ft^3/min)	23 ±2	21 ±2	26 ±2
Powder feed rate (rpm)	250-330	300-400	150-200
Spraying distance (mm.)	380	380	380
Temperature (°c)	3000-3600 °c	3000-3600 °c	3000-3600 °c

2.1 Microstructure and surface hardness of coatings

The micro-hardness of coating films were investigated by Vickers micro-hardness tester (Matsuzawa MM5321X model MMT-X7-LCD). This tester micro-hardness of coating films was found. Twenty

IOP Conf. Series: Materials Science and Engineering **361** (2018) 012002 doi:10.1088/1757-899X/361/1/012002

positions of measuring were the 10-testing points on flat surface and 10-testing points on cross section as shown in Figure 1. The measuring method used the applied loads of 100 g and time range of 10 sec. with each testing point. In addition, the microstructure analysis of specimens was used for the scanning electrode microscopy (SEM) and energy dispersive spectroscopy (EDS).

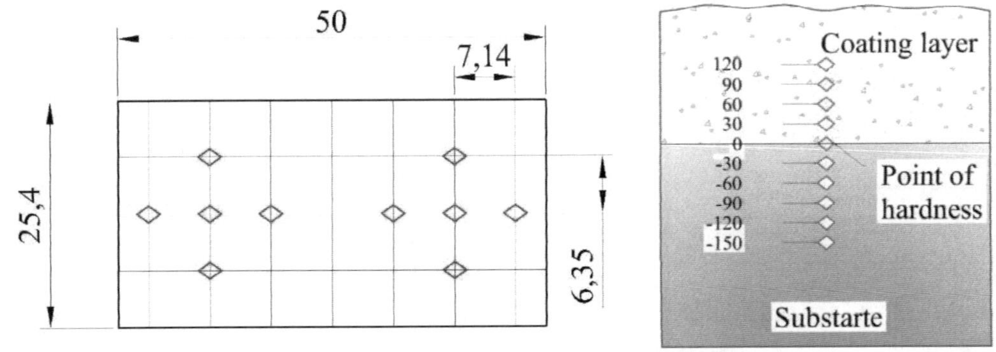

(a) Testing point on flat surface. (b) Testing point on a cross section.

Figure 1. Positioning of micro-hardness testing.

3. Result and Discussion

3.1 Microstructure

The SEM micrographs (as shown in Figure 2) were exhibited with the three coating films on cross section and deposition characteristic of coating films. These micrographs demonstrated that the coating films in good deposition with the specimens, Moreover, the coating films indicated a very homogeneous microstructure [5]. However, the porosity value in the coating films on cross section was low according to the evident. The EDS analysis of coating film on the flat surface was shown in Figure 3 and Table 4.. According to the SEM micrographs results, the powder particles consist of W, Cr, Ni, Co and C. The as sprayed tungsten based alloy composite results hard W particle results in the formation of WC during the remelting [6].

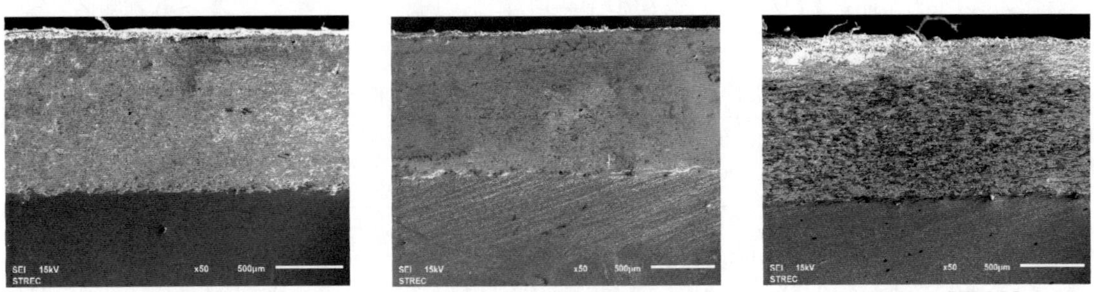

(a) AST 908 (b) AST 915 (c) AST 928

Figure 2. SEM micrographs of three coating films on cross section.

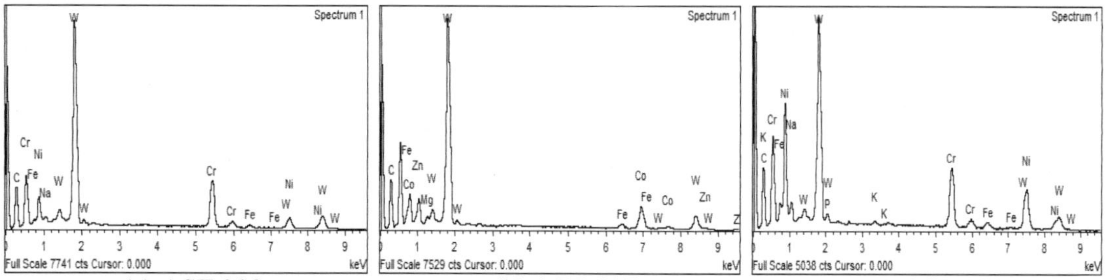

(a) AST 908 (b) AST 915 (c) AST 928

Figure 3. EDS analysis of coating films on flat surface.

Table 4. The EDS analysis of each coating films on flat surface.

Type of powder	Chemical composition (%)										
	C	Na	P	Zn	Co	Mg	K	Cr	Fe	Ni	W
AST 908	18.65	0.5	-	-	-	-	-	18.52	1.27	9.33	51.73
AST 915	22.81	-	-	4.36	16.72	0.67	-	-	2.6	-	52.84
AST 928	20.86	1.45	0.52	-	-	-	0.43	14.9	2.15	25.48	34.39

3.2 Vickers Micro-hardness

The variation of micro-hardness on coating films was shown in Figure 4. It revealed that AST 908 coating film higher hardness of 1583.9 HV was achieved whereas the hardness of specimen (AISI 1050 steel) was 300.5 HV. In addition, the AST 915 coating film and AST 928 coating film were improved the hardness of 1476.9 HV and 1316.4 HV respectively. It indicated that the hardness of specimen (81.30%) was lower than AST 908 coating film.

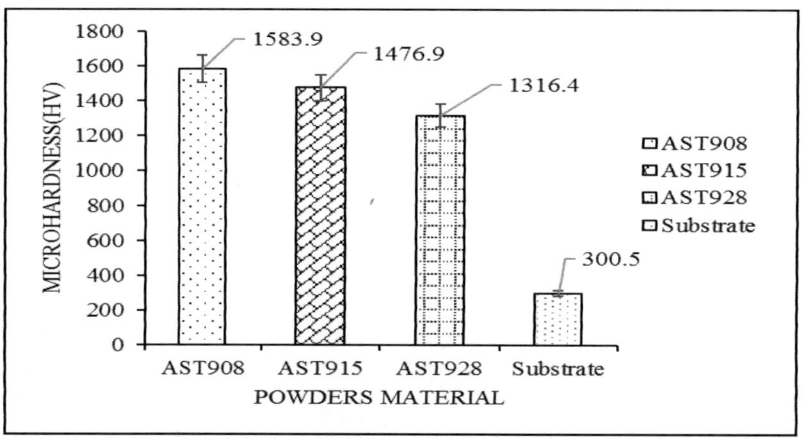

Figure 4. The variation of micro-hardness on coating films and substrate.

Figure 5, showed the micro-hardness along the cross section of substrate and coating films. It could be seen that the increase of the micro-hardness by adding the spray distance from the substrate to coating layer of AST 908 coating film, the surface coating of 1559.3 higher hardness HV by gradually pressed of the position shifts from surface coating towards bond layer. At the bond layer it is reduced to 699.6 HV and further reduced in substrate material to 313.4 HV. The AST 915 powder coating and AST 928 coating improved the micro-hardness on the AISI 1050 material by 1468.4 HV and 1306.2 HV respectively.

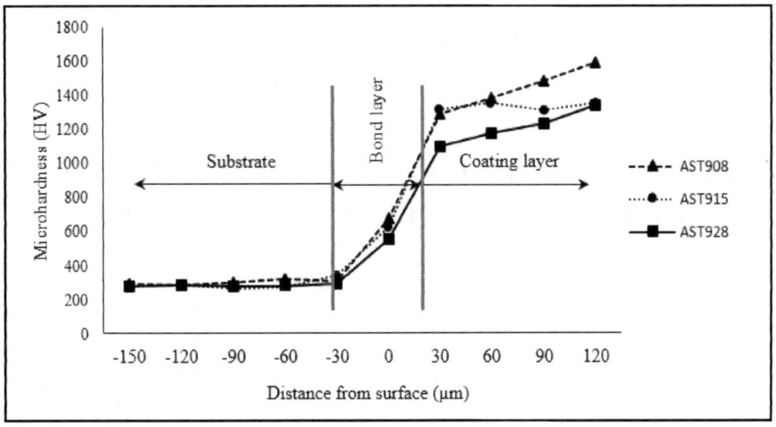

Figure 5. The spray distance of micro-hardness on coating films and substrate.

4. Conclusion

In this work, the surface characteristics during HVOF spayed coating on AISI 1050 substrate material using three different powder coatings have been investigated. The micro-hardness and microstructure were studied. The following conclusions can be drawn as follow:

- The cross section micrographs demonstrated that the coatings reasonably had good interface with the substrate material and the coating showed a very homogeneous microstructure.
- The micro-hardness of substrate was 81.30% lower than the micro-hardness of surface coating when using AST 908 powder particles.
- In the micro-hardness along the cross section, the AST 908, AST 915 and AST 928 powder coatings improved the micro-hardness on the AISI 1050 material by 1559.3 HV, 1468.4 HV and 1306.2 HV respectively.

References

[1] Andras M, Gabor B and Andras B 2013. Hardness test and microstructure analysis of NiCrBSi sprayed, Laser remelted coating. *Production Processes and Systems* Vol 6 (1) pp 35-46

[2] Ahmad N A, Kamdi Z, Mohamed Z, Omar A S, Latif N A and Mohd Tobi A L 2016. Charecterization of WC-10Ni HVOF Coating for Carbon Steel Blade. *IOP Conf. Series : Materials Science and Engineering* 165 012022

[3] Serkan I, Cihan O, Ozkan K, Mehmet A and Cihangir T S 2015. Microstructure and Hardness Properties of high velocity fuel (HVOF) Sprayed WCCo-SiC coating. *Kastamonu University Journal of Engineering and Sciences* 1(1) pp 1-7

[4] Raghu K R M R, Sarcar M M M and Ramanaiah N 2013. Tribological Behavior of WC-Co/NiCrAlY Coating on Ti-6Al-4V. *International Journal of Advanced Science and Technology* Vol 57

[5] Karagoz M, Islak S, Buytoz S and Kurt B 2011. microstructure Charecteristics of high velocity fuel (HVOF) Sprayed NiCrBSi-SiC Composite Coating on a Low Alloy Steel. *6th International Advanced Technologies Symposium (IATS 11)* 16-18 may 2011 Elazig Turkey

[6] Nitesh K J and Hariharan P 2012. Influence of Hardness By WC Based Coating On Alsi Alloy And Grey Cast Iron Using HVOF Coating Method. *IOSR Journal of Mechanical and Civil Engineering (IOSR-JMCE)* 2320-334x pp 36-42

Acknowledgement

The authors would like to thank Branch of Advanced Manufacturing Technology, Faculty of Engineering, for financial support of this research.

IOP Conf. Series: Materials Science and Engineering **361** (2018) 012003 doi:10.1088/1757-899X/361/1/012003

Influence of process parameters on surface roughness in single point incremental forming using dummy sheet

Vikas Sisodia, *Shailendra Kumar

Department of Mechanical Engineering, Sardar Vallabhbhai National Institute of Technology, Surat, India

*E-mail: skbudhwar@gmail.com

Abstract. Single point incremental sheet forming (SPIF) is an advanced sheet metal forming process. It is flexible and does not require dedicated punch and die set-up for forming operation. But due to its high surface roughness, it is still not so popular in sheet metal industries. But when SPIF is performed with dummy sheet at the top of target sheet, then this limitation is eliminated to some extent. In the present paper, research work involved in experimental study on influence of process parameters on surface roughness in SPIF using dummy sheet is described. Five process parameters namely dummy sheet thickness, tool size, step size, wall angle and feed rate are considered in the present study. From the analysis of variance (ANOVA), it is found that parameters namely dummy sheet thickness, tool size, step size, and wall angle are significant and feed rate is insignificant. The surface roughness decreases with increase in wall angle, dummy sheet thickness and tool size while increases with the increases in step size.

1. Introduction

Single point incremental sheet forming (SPIF) process is advanced sheet metal forming in which the requirement of dedicated punch and die setup is completely eliminated. The product is directly fabricated from CAD model itself, thus eliminates the need of fixture and punch-die setup. It offers higher flexibility as compared to conventional sheet metal forming. The process can be performed on any 3-axis CNC milling machine. But even though with such advantages, parts produced by SPIF has high surface roughness due to waviness caused by the forming tool. Due to this limitation, SPIF is not so popular in sheet metal industries.

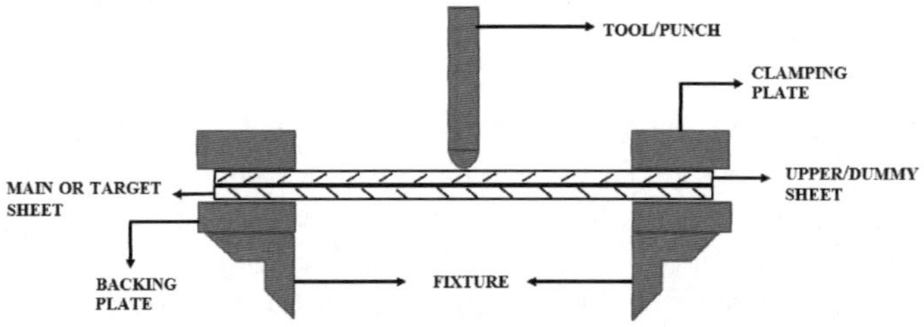

Figure 1. SPIF process using dummy sheet

However, as reported in literature when SPIF is performed with dummy sheet at the top of main or target sheet, the surface roughness is minimized to some extent. In this case all tool marks appear on

Content from this work may be used under the terms of the Creative Commons Attribution 3.0 licence. Any further distribution of this work must maintain attribution to the author(s) and the title of the work, journal citation and DOI.
Published under licence by IOP Publishing Ltd

dummy sheet while smooth surface is obtained at the main or target sheet. In SPIF process using dummy sheet, two sheets are deformed simultaneously. The main or target sheet is deformed by the virtual radius of the tool (virtual tool size) which is combination of tool diameter itself followed by the curve traced by tool on the upper sheet. Figure 1 shows SPIF process using dummy sheet on the top. Skjødt et al. [1] studied SPIF process using dummy sheet in order to overcome the limitation of poor surface finish in conventional SPIF process. It was reported that when DC-04 sheet was used as dummy sheet over AA1050-H111 grade aluminium the surface of main sheet appeared to be smooth also with this combination there was increase in achievable wall angle by 5 degrees. Martins et al. [2] presented the theoretical analysis about state of stress in SPIF process. They reported that the use of dummy sheet influences formability of main/target sheet material and limits maximum achievable forming angle or wall angle. This is due to the fact that the dummy sheet virtually increases diameter of tool. The increase of this virtual tool diameter leads to an increase of meridional stress, thereby, increasing hydrostatic stress which leads to reduced formability. Alves et al. [3] compared conventional SPIF of a single sheet and SPIF using dummy sheet with the help of experimental investigation. They used DC-01 steel sheet of 0.63 mm as dummy sheet over AA1050-H111grade aluminium. They reported that the use of dummy sheets can resolve problem of poor surface finish. The surface roughness value was very less in case of SPIF using dummy sheet as compared to the results obtained in conventional SPIF process. Also, the thickness distribution obtained after experimentation is close to the value obtained by sine law (only for higher wall angles). Silva et al. [4] used tailor friction stir welded blanks (TWB) of AA1050-H111 as sheet material for SPIF process with two different sheet thickness i.e. 1.5 mm and 2 mm. They used the dummy sheet over the welded blanks in order to protect welding seam from the rotating tool. They reported that very slight decrease in formability was observed along with good surface characteristics. A comparison is also done between conventional SPIF and SPIF using dummy sheet (Figure 2).

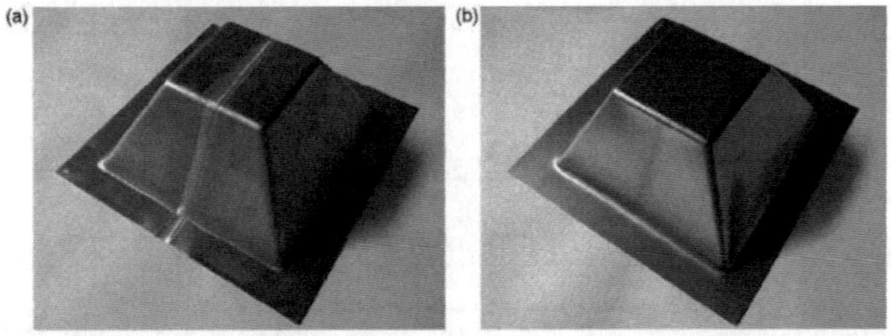

Figure 2. (a) Conventional SPIF of TWB of Aluminium AA1050-H111 with a uniform 2.0 mm thickness profile. (b) SPIF using dummy sheet [4]

Lasunon [5] investigated the influence of wall angle, step size and feed on surface roughness and reported that less step size and high wall angle is beneficial for good surface finish. Echrif and Hrairi [6] studied the influence of step size, tool size, spindle speed and feed on surface roughness and reported that bigger tool size and smaller step size give smooth surface. Influence of tool path, tool diameter and step size on the surface roughness (R_a) was studied by Jagtap et al. [7]. It was observed that with low step size and high tool diameter R_a value decreases. Najafabady and Ghaei [8] reported that with the increase in step size, surface roughness increases.

From the review of available literature, it is found that very less research efforts have been applied to study SPIF process using dummy sheet. Influence of process parameters on surface roughness of formed parts in case of SPIF process using dummy sheet has not been reported. The present study is focused on influence of process parameters of SPIF using dummy sheet on surface roughness of formed parts. Five process parameters namely dummy sheet thickness, tool size, step size, wall angle and feed rate are considered in the present study.

The subsequent sections of this paper highlight experimental plan, results and discussion and finally conclude the present study.

2. Experimental Plan

All experiments of present study are performed on a 3-axis milling machine (model – Dart of M/s Batliboi Ltd., Surat, India). Experimental set-up is shown in Figure 3. The fixture as mounted on machine table is fabricated. The objective of present work is to investigate the influence of five process parameters namely dummy sheet thickness, wall angle, step size, tool size and feed rate on surface roughness (R_a value) characteristics. The experiments are designed as per Box Behnken DOE. Total 46 experiments are performed. Figure 4 depicts the formed components. The set of parameters with their levels taken are listed in Table 1. Table 2 represents the parameters held constant during experimentation. Work piece material AA1050 is taken for experimentation both for dummy and target sheet.

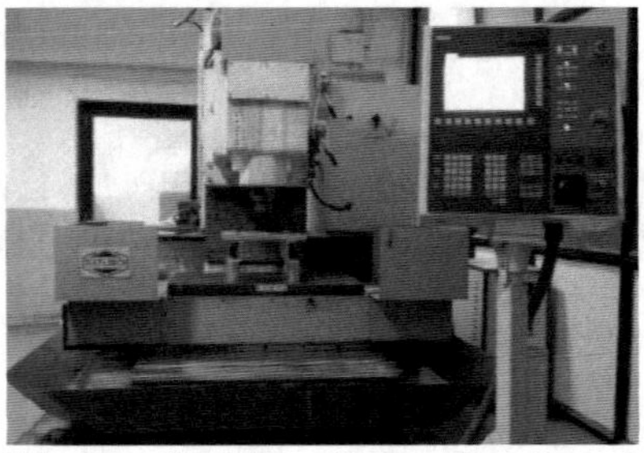

Figure 3. Experimental setup for SPIF using dummy sheet

Figure 4. Formed parts using SPIF with dummy sheet

Table 1. Process parameters and their level

S.No.	Parameters	Unit	Level 1	Level 2	Level 3
1	Wall angle	degree	35	50	65
2	Step size	mm	0.4	0.7	1
3	Tool size	mm	6	9	12
4	Feed rate	mm/min	1000	2000	3000
5	Dummy sheet thickness	mm	0.51	0.71	0.91

Table 2. Values of process parameter held constant

Parameter	Unit	Value
Tool path type	NA	Contoured
Spindle speed	rpm	0
Target Sheet thickness	mm	0.91
Mouth diameter of cone	mm	110
Tool material	NA	HSS

In present experimental study, 3 levels of dummy sheet thickness is considered in order to find out the influence of dummy sheet thickness on R_a value. The surface roughness (R_a) value is measured by using Mitutoyo surf-test SJ-310 in three distinct regions - bottom region (A), middle region (B) and top region (C). In each region eight readings are taken. The average of three regions is taken and represented as Average R_a. So total 24 reading are taken for each sample.

3. Results and Discussion

The R_a value is measured in three distinct regions and then average is taken. Table 3 lists the measured R_a values of experiments. The analysis of variance (ANOVA) is carried out to identify the significant variables and to quantify their effects on the response characteristics. It is necessary to find out the parameter that significantly affects the performance characteristics. The ANOVA table decomposes the variability of response characteristics into contribution due to various factors. Table 4 shows the ANOVA table for surface roughness (R_a).

Table 3. Measured Surface Roughness (R_a)

Exp. No.	Dummy sheet thickness (mm)	Tool size (mm)	Step size (mm)	Wall angle (degree)	Feed rate (mm/min)	R_a (μm)
1	0.51	6	0.7	50	2000	1.313
2	0.91	6	0.7	50	2000	1.059
3	0.51	12	0.7	50	2000	1.013
4	0.91	12	0.7	50	2000	0.983
5	0.71	9	0.4	35	2000	0.985
6	0.71	9	1	35	2000	1.309
7	0.71	9	0.4	65	2000	1.126
8	0.71	9	1	65	2000	1.019
9	0.71	6	0.7	50	1000	1.129
10	0.71	12	0.7	50	1000	1.121
11	0.71	6	0.7	50	3000	1.271
12	0.71	12	0.7	50	3000	1.038
13	0.51	9	0.4	50	2000	1.092
14	0.91	9	0.4	50	2000	1.132
15	0.51	9	1	50	2000	1.309
16	0.91	9	1	50	2000	1.094
17	0.71	9	0.7	35	1000	1.1
18	0.71	9	0.7	65	1000	1.047
19	0.71	9	0.7	35	3000	1.039
20	0.71	9	0.7	65	3000	1.052
21	0.71	6	0.4	50	2000	1.229
22	0.71	12	0.4	50	2000	1.065
23	0.71	6	1	50	2000	1.296
24	0.71	12	1	50	2000	1.1067
25	0.51	9	0.7	35	2000	1.0777
26	0.91	9	0.7	35	2000	0.973
27	0.51	9	0.7	65	2000	1.027
28	0.91	9	0.7	65	2000	0.9947
29	0.71	9	0.4	50	1000	1.057
30	0.71	9	1	50	1000	1.261
31	0.71	9	0.4	50	3000	1.1977
32	0.71	9	1	50	3000	1.064
33	0.51	9	0.7	50	1000	1.152
34	0.91	9	0.7	50	1000	0.964
35	0.51	9	0.7	50	3000	1.0153
36	0.91	9	0.7	50	3000	1.1047
37	0.71	6	0.7	35	2000	1.229
38	0.71	12	0.7	35	2000	1.0387
39	0.71	6	0.7	65	2000	1.0643
40	0.71	12	0.7	65	2000	1.0227
41	0.71	9	0.7	50	2000	1.154
42	0.71	9	0.7	50	2000	1.178
43	0.71	9	0.7	50	2000	1.224
44	0.71	9	0.7	50	2000	1.1927
45	0.71	9	0.7	50	2000	1.2157
46	0.71	9	0.7	50	2000	1.169

From ANOVA It is found that process parameters namely tool size, dummy sheet thickness, step size and wall angle are significant parameters. Feed rate is found to be insignificant parameter. As depicted in Figure 5, with the increase in tool size, R_a value decreases. This is because with higher tool size, more contact area is covered by tool on the sheet i.e. the area of contact increases. Also, with higher tool size the tool approximately covers 40% to 50% contact zone of the previous loop in the subsequent loop, thus supressing the scallop height.

Table 4. ANOVA Table for Surface Roughness

Source	Sum of Squares	df	Mean Square	F Value	p-value Prob > F	
Model	0.42	20	0.021	18.11	< 0.0001	significant
A-Dummy Sheet Thickness	0.030	1	0.030	26.24	< 0.0001	
B-Tool Size	0.090	1	0.090	78.62	< 0.0001	
C-Step Size	0.021	1	0.021	17.98	0.0003	
D-Wall Angle	9.935E-003	1	9.935E-003	8.65	0.0070	
E-Feed Rate	1.519E-004	1	1.519E-004	0.13	0.7192	
AB	0.013	1	0.013	10.92	0.0029	
AC	0.016	1	0.016	14.15	0.0009	
AE	0.019	1	0.019	16.74	0.0004	
BD	5.528E-003	1	5.528E-003	4.81	0.0378	
BE	0.013	1	0.013	11.02	0.0028	
CD	0.046	1	0.046	40.42	< 0.0001	
CE	0.029	1	0.029	24.81	< 0.0001	
A^2	0.046	1	0.046	40.17	< 0.0001	
D^2	0.070	1	0.070	60.61	< 0.0001	
E^2	0.019	1	0.019	16.74	0.0004	
Residual	0.029	25	1.149E-003			
Lack of Fit	0.025	20	1.251E-003	1.69	0.2930	not significant
Pure Error	3.698E-003	5	7.395E-004			
Cor Total	0.44	45				

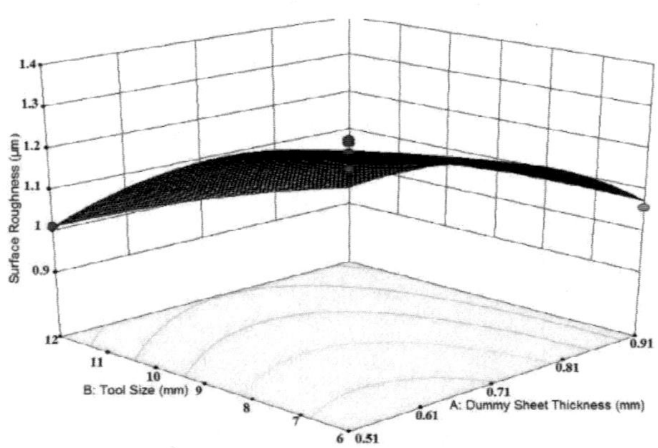

Figure 5. Influence of tool size and dummy sheet thickness on surface roughness

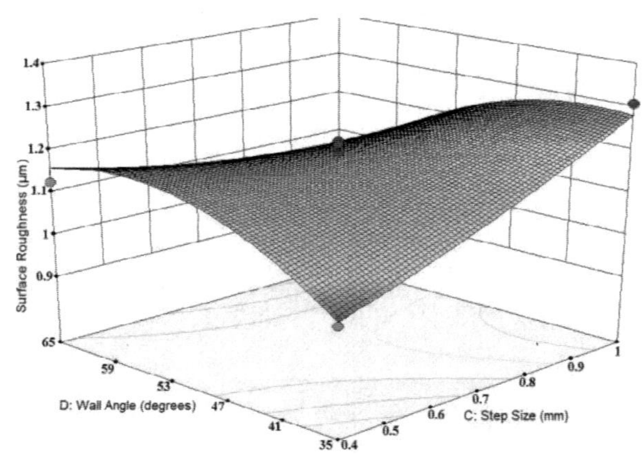

Figure 6. Influence of wall angle and step size on surface roughness

As the dummy sheet thickness increases, R_a value slightly increases and then decreases. The slight increase in R_a value is due to predominant effect of other parameters (like step size) than dummy sheet thickness itself. By using dummy sheet over target sheet, the virtual tool size (in terms of tool diameter) increases and as tool size increases the R_a value decreases. Figure 6 shows the influence of wall angle and step size on R_a value of target or main sheet. With the increase in step size, surface roughness (R_a value) increases. This is because surface waviness increases as step size increases. The reason for increased waviness of surface is due to increase in scallop height at larger step size [9]. Further, with the increase in wall angle, R_a value decreases. This is due to the fact that as wall angle increases the direct distance between the two subsequent loop decreases resulting in improved surface

IOP Conf. Series: Materials Science and Engineering **361** (2018) 012003 doi:10.1088/1757-899X/361/1/012003

finish. Feed rate has no influence on R_a value. This is because all the adverse effects like surface scratches and wear caused by high feed rate are completely faced by dummy sheet as also reported by Silva et al. [10]. Thus, good surface finish is obtained on target sheet. As feed rate is found insignificant in influencing surface roughness, therefore use of high feed is suggested to reduce forming time.

4. Conclusion

The influence of process parameters of SPIF process using dummy sheet on surface roughness of formed part has been investigated. Five parameters namely dummy sheet thickness, step size, tool size, wall angle and feed rate are considered. The findings of the present work are enumerated as under -

(i) Process parameters namely dummy sheet thickness, tool size, step size and wall angle are found significant in influencing R_a value of formed part (i.e. target sheet) and feed rate is insignificant.
(ii) There is considerable influence of dummy sheet thickness and wall angle on R_a value. With increase in dummy sheet thickness and wall angle, R_a value decreases.
(iii) As tool size (i.e. tool diameter) increases, R_a value decreases.
(iv) With increase in step size, R_a value increases.
(v) As the feed rate is found to be an insignificant parameter influencing R_a value, therefore high feed rate is suggested to reduce forming time.

The above findings are certainly useful in improving surface characteristics of formed part produced by SPIF using dummy sheet to make this process acceptable in sheet metal industries.

References

[1] Skjødt M, Silva B, Bay N, Martins P A and Lenau T 2007 *Proc. of 2nd Int. Conf. on New Forming* pp 267-76. BIAS Verlag.
[2] Martins P A, Bay N, Skjødt M and Silva M B 2008 *CIRP ANN. – Manuf. Techn.* 57(1) pp247-52.
[3] Alves M L, Silva M B, Alves L M and Martins P A 2008 *Proc. of Int. Conf. on Experimental Mechnics and 7th Asian Conference on Experimental Mechanics* pp73750N. International Society for Optics and Photonics.
[4] Silva M B, Skjødt M, Vilaca P, Bay N and Martins P A 2009 *J. Mater. Process. Technol.* 209(2) pp811-20.
[5] Lasunon O.U. 2013 *Adv. Mat. Res.* 753 pp203-06.
[6] Echrif S B M and Hrairi M 2014 *Mater. Manuf. Processes* 29(6) pp697–703.
[7] Jagtap R, Kashid S, Kumar S and Hussein H M A 2015 *Adv. Mater. Process. Technol.* 1(3-4) pp465-73.
[8] Najafabady S A and Ghaei A 2016 *Int. J. Adv. Manuf. Tech.* 87(9-12) pp3579-88.
[9] Aerens R, Eyckens P, Van-Bael A and Duflou J R 2010 *Int. J. Adv. Manuf. Tech.* 46(9) pp969-82.
[10] Silva P J, Leodido L M and Silva C R M 2013 *Key Eng. Mater.* 554 pp2285-92.

IOP Conf. Series: Materials Science and Engineering **361** (2018) 012004 doi:10.1088/1757-899X/361/1/012004

Effect of Si on Microstructure and Corrosion Behavior of CoCrMo Alloys

Chonlawit Peaubuapuan, Phacharaphon Tunthawiroon, Kannachai Kanlayasiri, Chumpol Yuangyai

Department of Industrial Engineering, Faculty of Engineering, King Mongkut's Institute of Technology Ladkrabang, Bangkok 10520, Thailand.

E-mail: chonlawit.fc@gmail.com

Abstract: The aim of this research was to investigate the effect of silicon (Si) on the microstructure and corrosion behavior of CoCrMo alloys. The concentration of Si added in CoCrMo alloys was 0.1, 0.5 and 1 wt %. The corrosion behavior of the present alloys was investigated using potentiodynamic polarization measurements. The normal saline solution with 0.9 wt % was used as an electrolyte. Polarization curves obtained from the polarization test were used to evaluate in terms of corrosion current density (i_{corr}), corrosion potential (E_{corr}) and corrosion parameters that be used to compute corrosion resistant property of CoCrMo alloys. The microstructure of a sample during the polarization test was compared using X-ray diffraction (XRD) and optical microscope (OM). The results indicated that the increase in Si slightly changed the microstructure of CoCrMo alloys and could enhance the resistance to corrosion of CoCrMo alloys in NaCl solution. In addition, the reduction of Cr and Mo concentration in CoCrMo alloys was found to be a significant influence on the decrease in corrosion resistance.

1. Introduction

The CoCrMo alloys are widely used in implants owing to their excellent mechanical properties, good biocompatibility and high corrosion resistance [1–3]. In addition, they are recommended for use in metal-on-metal joint replacement applications because of their high wear resistance and hardness [4,5]. However, CoCrMo alloys which expose to the environment is undergone an oxidation as well as other metallic biomaterials. As well-known, the oxidation is one of the corrosion degradation mode in which the oxide films are formed at the interface of the alloy and environment. In the oxidation process, although, spontaneous passive film forms, the loss of metals are simultaneously taken place. In general, it is accepted that the good corrosion resistance of CoCrMo alloys is attributed to the high concentration of Cr which is play an important role on the formation of an inert Cr_2O_3 film [6]. In contrast, Cr and Mo were reported to be a cause of the formation of intermetallic phase, named σ phase, which decreased the corrosion properties of these alloys [7]. Therefore, the concentration of Cr and Mo should be controlled to obtain the high performance of alloys. Si is widely used as an alloying element to enhance the corrosion resistance of steels and alloys in various conditions. In addition Si adding to CoCrW could retard the precipitation of the brittle σ phase [8]. However, no research related to the effect of Si on corrosion resistance of CoCrMo alloys has been carried out. Thus, in this study, three kinds of CoCrMo alloys with different Si concentrations (0.1, 0.5 and 1.0 wt %) are investigated

Content from this work may be used under the terms of the Creative Commons Attribution 3.0 licence. Any further distribution of this work must maintain attribution to the author(s) and the title of the work, journal citation and DOI.
Published under licence by IOP Publishing Ltd

the corrosion behavior in NaCl solution. For comparison, an alloy with reduced Cr and Mo is also performed the corrosion test in the same condition with the above-mentioned three alloys.

2. Materials and methods

Three kinds of Cobalt-Chromium-Molybdenum (CoCrMo) alloys with 0.1, 0.5 and 1 wt% Si and another alloy which has the lower Cr and Mo concentration were chosen for this study. Table 1 shows the chemical compositions of studied alloys. The all investigated alloys were cast using vacuum melting. In order to obtain the homogeneous structure, as-cast alloys were solution treated at 1250 °C for 24 hours followed by the thermo-mechanical process by hot forging at 1200 °C. Subsequently, alloys were rapidly quenched to room temperature in water. The disk samples were cut using wire electrical discharge machining (wire-EDM). The surface of sample was ground by silicon emery paper, polished with 1 and 0.3 μm alumina powder (AP-A suspension, Struers). Mirror-like surface was cleaned in alcohol ultrasonic cleaner for 10 minutes. Prior to corrosion tests, the phases identification of alloys used X-ray diffraction (XRD) analysis (Shimadzu labx xrd-6100) equipped with X-ray diffractometer (Cu Kα radiation source). The microstructure of the alloy was characterized by optical microscope (OM; ZEISS Axio lab).

Tabel 1. The chemical compositions of investigated CoCrMo alloys

Alloy composition (wt %)	Co	Cr	Mo	Si	Mn	C	N
Co-28Cr-6Mo-0.1Si	Bal.	27.5	6.1	0.156	0.614	0.056	0.135
Co-28Cr-6Mo-0.5Si	Bal.	28.1	6.15	0.466	0.521	0.054	0.148
Co-28Cr-6Mo-1.0Si	Bal.	28.2	6.17	0.966	0.593	0.054	0.128
Co-26Cr-5Mo-1.0Si	Bal.	26.5	5.5	1.0	0.75	0.065	0.100

The potentiodynamic polarization measurement was performed in three electrode systems including the working electrode (sample), Pt counter electrode and activated carbon reference electrode. The applied potentials ranging from -0.8 V to 0.4 V (SCE) were generated using a Potentiostst/Galvanostst PGSTAT 302N (Autolab, EcoChemie, Netherlands). The polarization experiments were performed with a scan rate of 1 mV/s in 0.9 wt % NaCl solution. The polarization test of each alloy was repeated at least three times to confirm the repeatability of experimental results. The corrosion rate was calculated using the parameters obtained by Nova 1.11 software (Electrochemistry software; Metrohm Autolab).

3. Results and Discussion

3.1. Phase identification and microstructure

The phase diagram of Co-29Cr-6Mo-(0-2wt%)Si was predicted by the Thermo-Calc V5.0 as shown in Fig. 1(a). Two types of crystal structures: face-centered cubic (γ-phase) and hexagonal close-packed (ε-phase) were displayed depend on the temperature. In addition, two intermetallic phases were shown in the calculated phase diagram; σ and μ. The results of the XRD analysis in Fig. 1(b), revealed that the microstructure of alloys was dominantly composed of a nearly single γ-phase. No peak related to the intermetallic phase was observed. This attributes to the homogenized temperature and cooling process which are favorable for the presence of a single γ-phase. In previous study, Li et al. has demonstrated that the dominant presence of the γ-phase provided a higher resistance to corrosion than that of ε-phase [9].

Figure 2 shows the initial microstructure of the investigated CoCrMo alloys, observed using the optical microscope. Equiaxed grains along with the small fraction of twin boundaries which is typical microstructure after deformation processes were observed for all alloys. Moreover, finer grains were observed when the concentration of Si in the CoCrMo alloys was increased. The finest grain size was obtained in the Co-26Cr-5Mo-1.0Si alloys. Some defects in the form of microspores which possibly remained from the casting process were detected in all alloys. The corrosion behavior of CoCrMo alloys with different grain size has been reported by Li et al [9] in that increasing of the quantity of grain boundaries would accelerated the anodic reaction resulting in the earlier occurrence of corrosion processes. To gain more details on the effects of different grain sizes causing by the Si addition, the corrosion behavior was performed and discussed in a next paragraph.

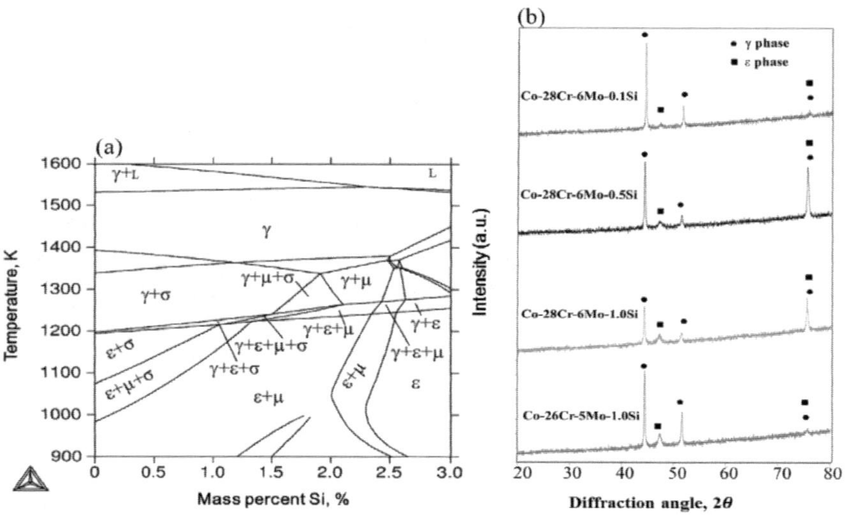

Figure 1. (a) Phase diagram of Co-29Cr-6Mo-*x*Si predicted by Thermo-Calc V5.0 and (b) XRD pattern of Co-Cr-Mo alloys with different Si contents.

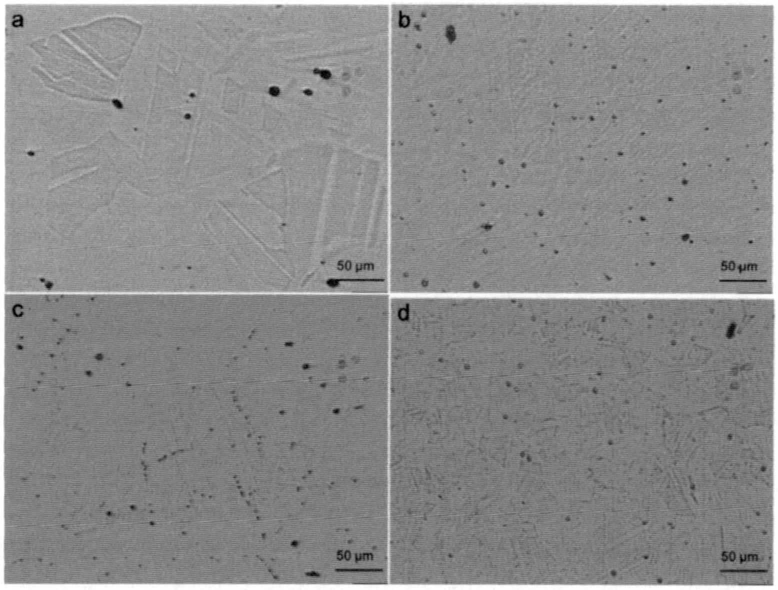

Figure 2. Optical micrograph of the investigated CoCrMo alloys: (a) Co-28Cr-6Mo-0.1Si, (b) Co-28Cr-6Mo-0.5Si, (c) Co-28Cr-6Mo-1.0Si and (d) Co-26Cr-5Mo-1.0Si.

3.2. Polarization testing

Figure 3 shows the polarization curves of Co-28Cr-6Mo-0.1Si, Co-28Cr-6Mo-0.5Si, Co-28Cr-6Mo-1.0Si and Co-26Cr-5Mo-1.0Si alloys in 0.9 wt % NaCl solution at room temperature. The shape of the curves exhibited a similar characteristic. The transition from cathodic to anodic regions of four alloys, generally referring to the open circuit potential (E_{ocp}), is located around the applied potentials ranging from -0.4 to -0.35 V (SCE). Among three alloys which have a similar Cr and Mo concentration, the

E_{ocp} of Co-28Cr-6Mo-1.0Si alloys was higher than that of Co-28Cr-6Mo-0.1Si and Co-28Cr-6Mo-0.5Si.

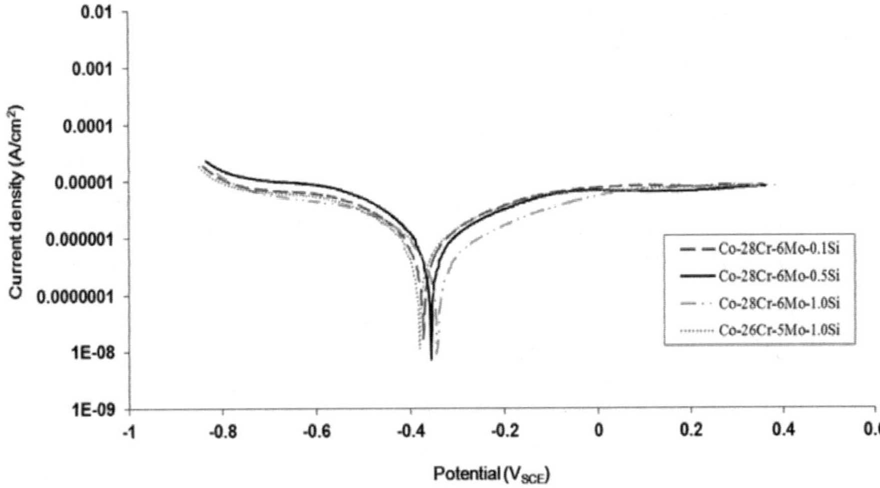

Figure 3. The polarization curves of CoCrMo with different Si alloys in 9 wt % NaCl solution.

This result indicates that the concentration of Si directly contributes to the enhancement of corrosion resistance although the higher concentration of Si reduce grain size of CoCrMo alloys. For the alloy with lower Cr and Mo concentration, the E_{ocp} was observed to be the most negative value, as seen in the polarization curve of Co-26Cr-5Mo-1.0Si in Fig. 3. This implies that reduced Cr and Mo leads to the decrease of corrosion resistance of alloys. This finding provides the interesting evidence on the influence of alloying elements on the corrosion resistance of CoCrMo alloys. Not only Cr/Mo but also Si plays a key factor in determining the corrosion behavior of investigated alloy series.

It has been reported that the high corrosion resistance of CoCrMo alloys is due to the formation of protective passive Cr_2O_3-rich film although Co is a primary constitutional element [10]. Mo also significantly enhances the localized corrosion resistance [11]. In addition, Si promoted the formation of a compact and single Cr_2O_3 passive film [12]. Thus, the alloys with higher Si concentration would have a higher ability to resist the corrosion from surrounding environment.

Table 2. Corrosion parameter of CoCrMo with different Si alloys in 9 wt % NaCl solution.

Alloy	Parameter				
	b_a (mV/dec)	b_c (mV/dec)	E_{corr} (mV vs SCE)	i_{corr} (μA/cm^2)	*Corrosion rate* (mm/year)
Co-28Cr-6Mo-0.1Si	0.5552	0.2551	-385.17	1.68	0.0195
Co-28Cr-6Mo-0.5Si	0.2060	0.2833	-332.47	0.65	0.0075
Co-28Cr-6Mo-1.0Si	0.2840	0.2626	-353.90	0.82	0.0095
Co-26Cr-5Mo-1.0Si	0.6544	0.3504	-386.05	1.51	0.0175

The parameters from the polarization curve are summarized in the Table 2. The corrosion rate in the unit of mm per year was calculated. It is clear that the corrosion rate is reduced from 0.0195 to 0.0095 mm/year for Co-28Cr-6Mo-0.1Si and Co-28Cr-6Mo-1.0Si, respectively while the reduced Cr and Mo alloys exhibit the highest corrosion rate. This observation is consistent with the previous observation on the corrosion property of CoCrMo alloys at high temperature application [13]. The presence of Si could slow down the reaction and dissolution of passive film in severe environment due to a positive effect from the oxide film formation. The further research about the surface characterization of these alloys is required to clarify with regarding to the significant effect of such a chemical composition of passive film.

IOP Conf. Series: Materials Science and Engineering **361** (2018) 012004 doi:10.1088/1757-899X/361/1/012004

4. Conclusions

The effect of Si on the microstructure and corrosion behavior of CoCrMo alloys was investigated. The finer microstructure of alloys was obtained by increasing of Si concentration. The corrosion behavior of CoCrMo alloys in 0.9 wt % NaCl solution exhibited a similar cathodic and anodic characteristic. The presence of Si in CoCrMo alloys could slightly enhance the corrosion resistance properties. The transition from cathodic to anodic region of low Si alloys took place much faster than that of high Si alloys. The amount of Cr and Mo also played a key role on the determination of corrosion behavior. Considering future direction of this work, corrosion behavior of these alloys should be characterized at near the human body temperature which might increase the rate of corrosion and oxidation. Furthermore, with further optimization of compositions that might provide the higher performance of corrosion resistance, Co-based alloys would become a potential materials for biomedical applications.

References

[1] Chiba A, Kumagai K, Nomura N and Miyakawa S 2007 Pin-on-disk wear behavior in a like-on-like configuration in a biological environment of high carbon cast and low carbon forged Co–29Cr–6Mo alloys *Acta Mater.* 55 1309–18

[2] Giacchi J and Morando C 2010 Microstructural characterization of as-cast biocompatible Co – Cr – Mo alloys Microstructural characterization of as-cast biocompatible Co – Cr – Mo alloys *Mater. Charact.* 62 53–61

[3] Bettini E, Eriksson T, Boström M, Leygraf C and Pan J 2011 Influence of metal carbides on dissolution behavior of biomedical CoCrMo alloy: SEM, TEM and AFM studies *Electrochim. Acta* 56 9413–9

[4] Chen Y, Li Y, Kurosu S, Yamanaka K, Tang N and Chiba A 2014 Effects of microstructures on the sliding behavior of hot-pressed CoCrMo alloys *Wear* 319 200–10

[5] Chen Y, Li Y, Kurosu S, Yamanaka K, Tang N, Koizumi Y and Chiba A 2014 Effects of sigma phase and carbide on the wear behavior of CoCrMo alloys in Hanks' solution *Wear* 310 51–62

[6] Zimmermann J and Colombi Ciacchi L 2010 Origin of the selective Cr oxidation in CoCr alloy surfaces *J. Phys. Chem. Lett.* 1 2343–8

[7] Kurosu S, Nomura N and Chiba A 2007 Effect of Sigma Phase in Co-29Cr-6Mo Alloy on Corrosion and Mechanical Properties *Adv. Mater. Res.* 26–28 777–80

[8] Yamanaka K, Mori M, Kuramoto K and Chiba A 2014 Development of new Co – Cr – W-based biomedical alloys: Effects of microalloying and thermomechanical processing on microstructures and mechanical properties *Mater. Des.* 55 987–98

[9] Zhang X, Li Y, Tang N, Onodera E and Chiba A 2014 Corrosion behaviour of CoCrMo alloys in 2 wt% sulphuric acid solution *Electrochim. Acta* 125 543–55

[10] Hodgson A W E, Kurz S, Virtanen S, Fervel V, Olsson C O A and Mischler S 2004 Passive and transpassive behaviour of CoCrMo in simulated biological solutions *Electrochim. Acta* 49 2167–78

[11] Yang B, Li J, Gong X, Nie Y and Li Y 2017 Effects of Cu addition on the corrosion behavior of NiCoCrMo alloys in neutral chloride solution *RSC Adv.* 7 40779–90

[12] Zhang X, Hu J, Ye L, Deng Y, Tang C, Yang L and Liu Z 2013 Effects of Si addition on microstructure and mechanical properties of Mg-8Gd-4Y-Nd-Zr alloy *Mater. Des.* 43 74–9

[13] Tunthawiroon P, Li Y and Chiba A 2015 Influences of alloyed Si on the corrosion resistance of Co (Tunthawiroon P)-Cr-Mo alloy to molten Al by iso-thermal oxidation in air *Corros. Sci.* 100

Acknowledgements

The author would like to thank Suppawat Chuvaree of Rajamangala University of Technology Krungthep for assistance in sample preparation.

IOP Conf. Series: Materials Science and Engineering **361** (2018) 012005 doi:10.1088/1757-899X/361/1/012005

Optimizing the resistance spot-welding process for dissimilar stainless steels

T Khuenkaew and K Kanlayasiri

Department of Industrial Engineering, Faculty of Engineering,
King Mongkut's Institute of Technology Ladkrabang, Bangkok 10520, Thailand

Email: tankhun_k@hotmail.com

Abstract. The objective of this study was to identify the ideal welding circumstances to conduct resistance spot-welding in the case of 316 austenite and 425 ferrite stainless steels, which offer the challenge of being dissimilar. The study examines a number of welding variables, including weld current, weld time, hold time, electrode force, and the upper and lower electrode types. Initially, a 2^{6-1} fractional factorial design was used for the screening of variables involved in the process. After this stage, the Box-Behnken design was employed for analysis of those variables and in order to specify the multiple response optimizations applicable to the size of nuggets through examination of the desirability function. The experiment was run using a resistance spot-welding machine at 50 kVA, 50 Hz. The quality of welded specimens was determined by their penetration, nugget diameter and nugget area. The optimal welding conditions were found to be 3.3 kN of electrode force, 25 cycles of weld time, 10,000 amperes of weld current, 50 cycles of hold time, and both R30 types of up-low electrodes. The optimal welding conditions provided nugget sizes of quality which were able to pass the specifications of the customer.

1. Introduction

The use of RSW (Resistance Spot Welding) has become commonplace in many industries where sheet metal must be joined. The technique is found in the manufacturing, automotive, and medical sectors due to its simplicity and relatively low cost [1-5]. RSW of stainless steel is an efficient joining process for the assembly of vehicle exhaust parts. Ferritic and austenitic stainless steel grades have been widely used in joint exhaust pipe parts due to its high corrosion resistance, decorative appearance, and excellence welding ability. RSW can be very complex when it involves stainless steel, since a number of factors must interact, most notably the thermal, electrical, mechanical, and metallurgical elements. The pieces of metal which are to be welded together are first placed in between the up-low electrodes prior to pressurization and the introduction of the electric current. Weld strength and quality is directly linked to the size and shape of the weld nuggets, whereby the nuggets of quality size can be produced simply by following the guidance offered with regard to suitable measurements in the manual produced by the Welder Manufacturers' Association. It is thus vital to choose suitable welding parameters for the welding process in order to achieve the optimal nugget size. However, the chosen welding parameters are not assured of being optimal [6].

Process optimization in RSW has already been studied for resistance spot-welding parameters in terms of nugget size. Muhammad, N et al. presented the use of a multi–objective center composite design and the Taguchi approach tested a design method which put forward the response surface

Content from this work may be used under the terms of the Creative Commons Attribution 3.0 licence. Any further distribution of this work must maintain attribution to the author(s) and the title of the work, journal citation and DOI.
Published under licence by IOP Publishing Ltd

method for characteristics with multiple qualities, which in this case would be the weld nugget radius or the HAZ width [7]. On the other hand, some research has shown multi-response optimizations (more than 2 responses). Ruggiero, A et al. investigated the use of the Box-Behnken experiment design method with 3 factors for 5 responses to a laser-welding process. It was necessary to perform this step to establish the main parameters for the process which can affect the weld bead profile for joints which have been welded [8]. Ragavendran, M et al. reported on the investigation of optimized weld parameters on the weldments of a hybrid laser TIG welded 316LN stainless steel sheet using a center composite design method with 4 factors for 3 responses [9].

The objective of this paper was to determine the optimal welding conditions for quality characteristics, which are penetration, nugget diameter and nugget area, on the RSW of dissimilar stainless steel grade 316 and grade 425. The variables used in this study relating to the welding conditions were weld current, weld time, electrode force, hold time, and upper and lower electrode types. 2^{6-1} fractional factorial design method and ANOVA were used to screen the RSW parameter factors on the nugget shape of RSW. Then, the Box-Behnken experiment design method for optimization of the experiment using RSM (Response Surface Methodology) and a desirability function were employed as multiple-response optimization tools. The tensile shear load and microstructure of the weld sample, which was welded using the optimal conditions, was evaluated.

2. Experimental procedure

2.1. Materials and Methodologies

The workpiece material used in this study had a thickness of 1.5 mm. The base materials used were austenitic stainless steel 316 (SUS316L) and ferritic stainless steel 425 (SUS425) sheets. Both materials were cut into plates of 125 x 45 x 1.5 mm in size, which were lap joined according to JIS Z3139 standard and spot welded using a 50 kVA, 50 Hz RSW machine [10]. Six welding factors were selected for experimentation. The weld nugget attributes of all the samples were measured to analyze their metallographic cross-sections in the area to be welded, as Figure 1 indicates.

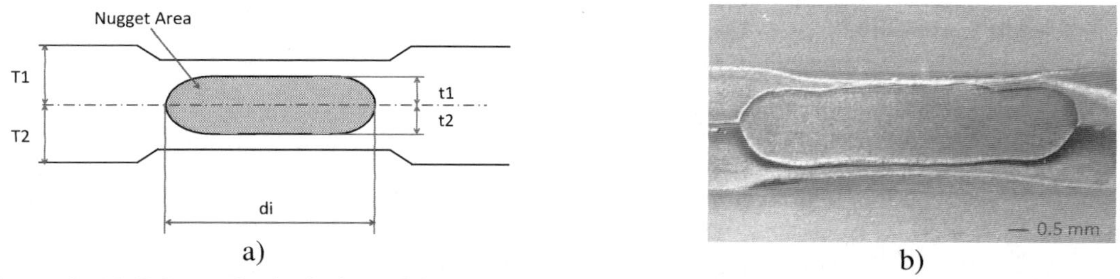

Figure 1. (a) Schematic depiction of the welding area; (b) Macrostructure of cross section weld sample.

2.2. Fractional factorial and optimized experiment design

Fractional factorials are widely used in screening experimentation, so in this study the design was of a general 2^{6-1} fractional factorial type with an additional 3-center point. Table 1 presents the various levels and factors applied. This experiment was created to identify the key welding factor among those in the response surface study. The data were then subjected to ANOVA analysis using a confidence level of 95%. RSM is a combination of a number of multiple regression methods and experimental designs. It can be applied to analyze the various factors which can stimulate the response [11]. The Box-Behnken design was used to optimize the main factors which can influence the development of the weld quality via the fractional factorial design.

Table 1. Experimental design factors and levels.

Factors	Symbol	2^{6-1} fractional factorial		Box Behnken.		
		Low level (-1)	Up level (+1)	Low level (-1)	Middle level (0)	Up level (+1)
Force(kN)	f	2	4	2	3	4
Welding current(Amperes)	I	7,000	10,000	7,000	8,500	10,000
Welding Time(Cycle)	Tw	25	50	25	38	50
Holding time(Cycle)	Th	0	50	0	25	50
Upper Electrode	Up ET	0	30	0	16	30
Lower Electrode	Low ET	0	30	0	16	30

Applying MINITAB statistics software, the experiment data were assessed using RSM in order to optimize the various response variables. The analysis makes it necessary to find a possible approximation for the true relationship linking the response surface and the independent variable. It is common to apply a second order model when analyzing response surfaces.

$$y = \beta_0 + \sum_{i=1}^{n} \beta_i x_i + \sum_{i=1}^{n} \beta_{ii} x_i^2 + \sum_{i<j} \sum \beta_{ij} x_i x_j + \varepsilon \quad (1)$$

in which β_0 denotes the constant, β_i, β_{ii} and β_{ij} are the respective coefficients of the linear, quadratic, and interaction variables; y denotes the response, or dependent variable; x_i and x_j denote the independent variables, while the error term is given by ε represents the influence of the parameters which are not included. The coefficient of determination R^2 indicated the fitting quality for the equation of the polynomial model.

The software can be used to measure the importance of each individual response and give a suitable weighting. The basic method is to begin by changing each of the responses given by y_i to become a desirability function d_i which falls somewhere within the range of 0 - 1, in which a greater value for d_i shows the greater desirability of the response value y_i. In the case when $d_i = 1$, this indicates a response which is wholly desirable [12]. In this study, the evaluation of each of the individual desirability responses, d_i was achieved through using the target for the responses y_i at the maximum value equation. Equation (2) reveals the overall objective function in conditions, where the changing degree of importance is given to the different response.

$$D = (d_1 {}^{x} d_2 {}^{x} d_3 {}^{x} \ldots\ldots {}^{x} d_n)^{1/n} \quad (2)$$

in which n represents the responses.

3. Results and discussion

3.1. 2^{6-1} Fractional factorial experiment results

The significance of each of the factors was assessed using normal distribution based on the standardized effect with a confidence level of 95%. The normal probability plot of standardize effect for each response observed that the main effect are the most important factors affecting the welding quality (t1, t2, di, Area). The R^2 values of the model were 99.64%, 99.95%, 99.99%, and 100%, respectively. The adjusted R^2 values of the model were 93.89%, 99.08%, 99.85%, and 99.99%, respectively. All of the factors were selected for optimization of the experiment.

3.2. Optimized experiment results

MINITAB statistical software was used to create and analyze the experiment data. The data from the experiments were subjected to multiple regression analysis applying a polynomial model of the second order. The significance for each regression coefficient was evaluated using the p-value. In this case the p-value was smaller than the confidence level of 0.05, indicating the statistical significance of the model. The R^2 values of this analysis were 88.17%, 86.61%, 91.26% and 93.09%, respectively, while the adjusted R^2 values of the model were 75.89%, 72.71%, 82.18% and 85.91%, respectively. These values were relatively moderate to high. For each of the responses, the data fit well to the model developed. The following second-order polynomial equation was used to explain t1, t2, di, Area removal efficiency, with the final reduced mathematical model in terms of both code and uncode factors determined by MINITAB software. The uncode mathematical model is shown below as Equations 3-6. The equations in terms of uncode factors can be given as:

$$\text{Penetration 1 (t1)} = 1.68926 + 0.00046(Tw) - 0.01132(I) - 0.00963(Th) - 0.02703(UpET) + 0.00454(Low\ ET) + 0.00006(Th)^2 + 0.00055(UpET)^2 - 0.00018(Tw*LowET) + 0.00009(I*Th) + 0.00016(I*UpET) \quad (3)$$

$$\text{Penetration 2 (t2)} = 3.1235 - 0.55063(f) - 0.032450(I) - 0.00607(UpET) - 0.02602(LowET) + 0.06678(f)^2 + 0.00020(I)^2 + 0.00010(Th)^2 + 0.00052(LowET)^2 + 0.00391(f*UpET) + 0.00018(I*LowET) - 0.00042(UpET*LowET) \quad (4)$$

$$\text{Nugget Diameter (di)} = 12.3727 - 2.7227(f) - 0.1460(Tw) - 0.0149(I) + 0.0321(UpET) + 0.0555(LowET) + 0.0016(Tw)^2 + 0.0021(LowET)^2 + 0.0189(f*Tw) + 0.0199(f*I) - 0.0014(I*LowET) - 0.0024(UpET*LowET) \quad (5)$$

$$\text{Nugget Area (Area)} = 34.4043 - 8.7522(f) - 0.2320(Tw) - 0.2226(I) - 0.1093(Tw) - 0.1626(UpET) - 0.0547(LowET) + 0.0049(Tw)^2 + 0.0008(Th)^2 + 0.0045(UpET)^2 + 0.0075(LowET)^2 + 0.0411(f*Tw) + 0.0541(f*I) + 0.0025(I*UpET) - 0.0074(UpET*LowET) \quad (6)$$

The optimization criteria employed in this study is shown in Table 2. The optimized graphics were carried out by response optimization plot and overlay response contour plot. The optimal values for RSW parameters were 3.3 kN of electrode force, 25 cycles of weld time, 10,000 amperes of weld current, 50 cycles of hold time and both R30 types of up-low electrodes. Response optimization results are for achieving 1.20 mm of penetration 1 (t1), 1.05 mm of penetration 2 (t2), 8.12 mm nugget diameter (di) and 16.80 mm^2 nugget area (Area). Optimization desirability was evaluated to be 1, showing that the parameters were all within acceptable working bounds.

Table 2. Optimization criteria used in study.

Condition	Goal	Lower	Target	Weight	Importance
Penetration1(t1)	Maximum	0.30	0.375	1	1
Penetration2(t2)	Maximum	0.30	0.375	1	1
Nugget diameter(di)	Maximum	3.674	4.899	1	1
Nugget Area(Area)	Maximum	8.787	10.435	1	1

3.3. Test for confirmation

The experiment required for confirmation involved testing using a particular mix of factors which had already been assessed earlier. The calculation results of the object parameters based on the optimum design parameters set are listed in Table 3. The percentage error for penetration 1, penetration 2, nugget diameter, and nugget area were 3.33%, 0.95%, 2.59% and 3.87%, respectively. The small percentage error demonstrates the efficiency of the optimization method.

In order to estimate the weldability of RSW, tensile strength was measured by the tensile shear load test. The specimens were prepared according to JIS Z3136 [13]. The tensile shear load for material sheet thickness was 1.5 mm, as recommended by the automotive industry. JIS Z3140 is greater than

7.65 kN [14]. The results indicated that most combinations were satisfactory with the criterion, assuring the reliability of the RSW of stainless steel. The tensile shear load values for specimens 1, 2, 3 were 14.76kN, 14.81kN and 14.51kN, respectively, meaning they achieved the JIS Z3140 standard.

Table3. Comparison of predicted and validation test results.

Response	Predicted	Desirability	Experimental				Error
			No.1	No.2	No.3	Average	(%)
Penetration(t1)	1.20	1	1.17	1.15	1.15	1.16	3.33%
Penetration(t2)	1.05	1	1.06	1.07	1.05	1.06	0.95%
Nugget diameter(di)	8.12	1	8.33	8.48	8.18	8.33	2.59%
Nugget area(Area)	16.80	1	16.14	16.26	16.05	16.15	3.87%

The cross section microstructure of welded specimens joined at optimal welding parameters was investigated by optical microscopic analysis. The type of microstructure of RSW for stainless steel considers two distinct regions, base metal and HAZ (Heat Affect Zone), as shown in Figure 2. Figures 2a, 2b, 2c, and 2d show the microstructure of weld nuggets and HAZ. The grains sizes are greater than the base metal and elongate parallel to the electrode force direction. Figures 2e and 2f illustrate the fusion zone region and also as the ferritic microstructure with various morphologies.

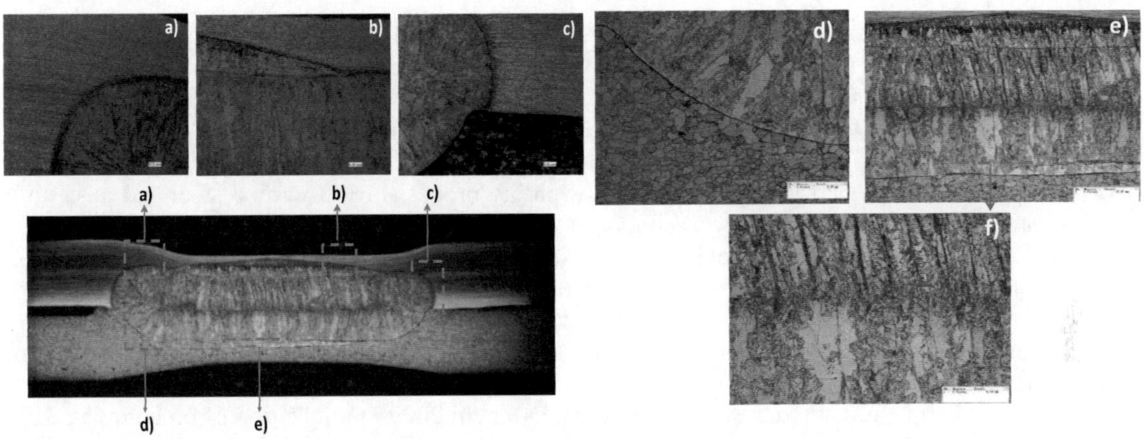

Figure2. Microstructure of cross section weldment sample (optimal welding parameters).

4. Conclusions

In the present paper, multi-response optimizations of RSW process parameters for dissimilar stainless steel grade 316 and grade 425 were successfully implemented to influence the quality of nugget size. Six variables were employed as in experiment. The 2^{6-1} fractional factorial design technique was applied in screening the parameter factors for RSW related to the nugget shape. Afterwards, the study used the Box-Behnken design for the investigation, while optimization was sought through the use of the desirability function. The optimal values of RSW parameters were 3.3 kN of electrode force, 25 cycles of weld time, 10,000 amperes of weld current, 50 cycles of hold time and both R30 types of up-low electrodes. Under optimal conditions, the tensile shear load of specimens achieved the JIS Z3140 standard.

References

[1] P.S. Wei and T.H. Wu 2014 Workpiece property effects on nugget microstructure determined by heat transfer and solidification rate during resistance spot welding *International Journal of Thermal Sciences* Vol 86 pp 421–429.

[2] Jae Hyung Kim, Yongjoon Cho and Yong Hoon Jang 2013 Estimation of the weldability of single-sided resistance spot welding *Journal of Manufacturing Systems* Vol 32 pp505– 512.

[3] Norasiah Muhammad, and Yupiter HP Manurung 2012 Design Parameters Selection and Optimization of Weld Zone Development in Resistance SpotWelding *World Academy of Science, Engineering and Technology* Vol 06 pp1184-1189.

[4] Yong Bing Li, Ze Yu Wei, Ya Ting Li, Qi Shen and Zhong Qin Lin 2013 Effects of cone angle of truncated electrode on heat and mass transfer in resistance spot welding *International Journal of Heat and Mass Transfer* Vol 65 pp400–408.

[5] Hamid Eisazadeh, Mohsen Hamedi and Ayob Halvaee 2010 New parametric study of nugget size in resistance spot welding process using finite element method *Materials and Design* Vol 31 pp149–157.

[6] Norasiah Muhammad, Yupiter HP Manurung, Mohammad Hafidzi, Sunhaji Kiyai Abas, Ghalib Tham and M. Ridzwan Abd. Rahim 2012 A Quality Improvement Approach for Resistance Spot Welding using Multi-objective Taguchi Method and Response Surface Methodology *International Journal on Advanced Science Engineering Information Technology* Vol 02 pp17-22.

[7] Norasiah Muhammad, Yupiter HP Manurung, Mohammad Hafidzi, Sunhaji Kiyai Abas, Ghalib Tham and Esa Haruman 2012 Optimization and modelling of spot welding parameters with simultaneous multiple response consideration using multi-objective Taguchi method and RSM *Journal of Mechanical Science and Technology* Vol 26 pp2365-2370.

[8] A. Ruggiero, L. Tricarico, A.G. Olabi and K.Y. Benyounis 2011 Weld-bead profile and costs optimisation of the CO2 dissimilar laser welding process of low carbon steel and austenitic steel AISI316 *Optics & Laser Technology* Vol 43 pp82-90.

[9] M. Ragavendran, N. Chandrasekhar, R. Ravikumar, Rajesh Saxena, M. Vasudevan and A.K. Bhaduri 2017 Optimization of hybrid laser – TIG welding of 316LN steel using response surface methodology (RSM) *Optics and Lasers in Engineering* Vol 94 pp27-36.

[10] Japanese Standard Association 1995 Method of Macro test for section of spot weld joint *Japanese Industrial Standard JIS Z3139* pp658-661.

[11] Praveen Sharma, Lakhvinder Singh and Neeraj Dilbaghi 2009 Optimization of process variables for decolourization of Disperse Yellow 211 by Bacillus subtilis using Box–Behnken design *Journal of Hazardous Materials* Vol 164 pp1024- 1029.

[12] Kannachai Kanlayasiri and Prajak Jattakul 2013 Simultaneous optimization of dimensional accuracy and surface roughness for finishing cut of wire-EDMed K460 tool steel *Precision Engineering* Vol 37 pp556– 561.

[13] Japanese Standard Association 1995 Method of tension test for spot-weld joint *Japanese Industrial Standard JIS Z3136* pp637-639.

[14] Japanese Standard Association 1989 Method of Inspection for spot welds *Japanese Industrial Standard JIS Z3140* pp814-821.

Acknowledgements

The researchers are grateful to Yokoyama Kogyo Thailand Co., Ltd for providing access to the RSW machine. The authors would also like to express their thanks to the Faculty of Engineering of King Mongkut's Institute of Technology Ladkrabang who offered support in covering the costs incurred in conducting this research.

IOP Conf. Series: Materials Science and Engineering **361** (2018) 012006 doi:10.1088/1757-899X/361/1/012006

Pressure-Controlled Dynamic Equation for the Phase-Field Crystal Method

J Em-Udom and N Pisutha-Arnond[1,2]

Department of Industrial Engineering, Faculty of Engineering, King Mongkut's Institute of Technology Ladkrabang, Bangkok 10520, Thailand

E-mail: [1]nirand.pi@kmitl.ac.th

Abstract. The phase-field crystal (PFC) method is a density-functional-type materials model with atomic resolution on diffusive timescale, and has been shown to be a robust tool for modeling materials phenomena. In this study, a pressure-controlled dynamic equation for the PFC model was developed to describe a system under externally applied hydrostatic pressure. The formulation is based on the established frameworks including the thermodynamics of solids, kinematics of deformation, and non-equilibrium thermodynamics. The proposed equation exhibits the tensile and compressive deformation rates that are different from those from the recently-proposed pressure-controlled dynamic equation. The derivation in this work also provides a framework that can be readily extended to describe a more complex system such as multi-component solid under non-hydrostatic stress.

1. Introduction

With increasing computing performance and availability, computational materials study has become a promising alternative to experimental study in terms of expanding our understanding of materials behavior and providing insight into new phenomena. Among wide ranges of available computational materials techniques, the phase-field crystal (PFC) method [1], has shown to be a robust materials modeling tool. The appeal of the PFC method is that the model exhibits atomic-scale resolution while operating in diffusive timescale, which has advantages over the conventional phase field model in terms of accuracy and self-consistency, and over other atomistic models in terms of computational expense. Such advantages result in numerous applications of the PFC method in describing materials phenomena such as liquid-solid transition, elasticity, plasticity, dislocation dynamics, and crack propagation (see Ref. [2] for comprehensive review in this topic).

The original formulation of the PFC method describes non-equilibrium behavior of a system through the minimization of the free energy under controlled temperature, mass, and volume. Due to the controlled-volume condition, the pressure is naturally not independent and cannot be directly controlled. Although this controlled-volume condition is applicable to many situations, it is often more experimentally relevant to simulate materials behavior under a pressure-controlled condition. Recently, Kocher and Provatas [3] proposed the scheme where the pressure-controlled dynamic equation is added to the PFC formulation to control the (internal) pressure through the specified external pressure. Although this equation can be successfully used to obtain equilibrium

[2] Corresponding author

Content from this work may be used under the terms of the Creative Commons Attribution 3.0 licence. Any further distribution of this work must maintain attribution to the author(s) and the title of the work, journal citation and DOI.
Published under licence by IOP Publishing Ltd

states of a system, the validity of the non-equilibrium behavior from this equation needs to be investigated due to certain postulations that might not be fully consistent with the established non-equilibrium thermodynamic framework.

Therefore, in this work, we proposed an alternative pressure-controlled dynamic equation that is consistent with the non-equilibrium thermodynamic framework. The equation is based on the thermodynamic description of solids by Larché and Cahn [4], the balance laws and kinematics of deformation from the work by Mishin et al. [5], the balance law from the Reynolds transport theorem [6, 7], and the linear relations between driving forces and fluxes from the classical irreversible thermodynamics [7, 8]. The result shows that the equation proposed in this work exhibits higher (lower) deformation rate under tensile (compressive) deformation than that exhibited by the equation by Kocher and Provatas [3]. Moreover, the derivation presented in this work provides a framework that can be readily extended to describe a more complex situation such as a multi-component system under non-hydrostatic stress.

The paper is organized as follows. In section 2, we provide background information on the PFC method and the pressure-controlled dynamic equation recently proposed by Kocher and Provatas [3]. In section 3, we present the derivation of the alternative pressure-controlled dynamic equation and show the comparison between the proposed and the original equations in Section 4. In Section 5, we provide a summary of our work.

2. PFC Method

The central part of the PFC method is the expression of the Helmholtz free energy which is written in terms of dimensionless quantities [9]:

$$F = \int_V f d\vec{r}\ ,\ f = \frac{\rho}{2}\left(-\epsilon + \left(1+\nabla^2\right)^2\right)\rho + \frac{\rho^4}{4}, \tag{1}$$

where f is the free energy density, ρ is the atomic number density field, ϵ is the model parameter related to temperature. The evolution of ρ is driven by dissipative dynamics through the conserved Cahn-Hilliard equation:

$$\frac{\partial \rho}{\partial t} = \nabla^2 \mu \equiv \nabla^2\left(\frac{\delta F}{\delta \rho}\right), \tag{2}$$

where $\mu = \delta F/\delta\rho$ is the chemical potential and can be obtained from the variational derivative of F with respect to ρ. The evolution equation (2) evolves the system along the free-energy-minimizing path under fixed temperature, mass and volume (or computational domain). This indicates that the (internal) pressure, the conjugate variable to the volume, is not independent. To provide a more direct control over the internal pressure, Kocher and Provatas [3] proposed an evolution equation for the volume or the pressure-controlled dynamic equation:

$$\frac{dV}{dt} = M\left(\bar{P}_{\text{int}} - \bar{P}_{\text{ext}}\right), \tag{3}$$

where $\bar{P}_{\text{int}} = (1/V)\int_V(-f+\mu\rho)\,d\vec{r}$ is the averaged internal pressure, $\bar{P}_{\text{ext}}$ is the externally applied pressure, and M is the mobility. The above equation provides a driving force for the system to deform (change in V) to minimize $\bar{P}_{\text{int}} - \bar{P}_{\text{ext}} \equiv \Delta\bar{P}$. If $\Delta\bar{P} > 0$, the system undergoes tensile deformation while $\Delta\bar{P} < 0$ results in compressive deformation.

3. Derivation of the Alternative Pressure-Controlled Dynamic Equation

3.1. Thermodynamics of Solids

We first introduce the thermodynamic description of solids proposed by Larché and Cahn [4] which considers the solids as networks of lattices occupied by atomic species and vacancies.

For a hydrostatically-stressed, single-component solid, the number of lattices sites (N_L) can be written as

$$N_L = N_A + N_v, \tag{4}$$

where N_A and N_v are the numbers of lattices occupied by the atomic species and vacancies, respectively. The internal energy, E, is then postulated to have the following functional dependence: $E(S, V, N_A, N_v)$, where S is the entropy and V is the volume. We can then write the differential of E as [10]

$$dE = TdS - PdV + \mu_A dN_A + \mu_v dN_v, \tag{5}$$

where T is temperature, μ_A and μ_v are the chemical potentials. Next, we consider the energy density $e_{V'} = E/V'$, where V' is the volume at the undeformed state. It can be shown using the standard thermodynamic relationships such as the Euler relation and Gibbs-Duhem equation that [10]

$$de_{V'} = Tds_{V'} - PdJ + \mu_{Av} d\rho_{A'}, \tag{6}$$

where $s_{V'} = S/V'$, $\rho_{A'} = N_A/V'$, $J = V/V'$, and $\mu_{Av} = \mu_A - \mu_v$. In arriving at the expression for $de_{V'}$, we use the assumption that the network is conserved or $\rho_{A'} + \rho_{v'} = \rho_{L'} \equiv$ Constant, where $\rho_{v'} = N_v/V'$ and $\rho_{L'} = N_L/V'$. Alternative to $e_{V'}$ which is defined by the undeformed volume (Lagrangian frame), we can define $e_V = E/V = e_{V'}/J$ which is the energy density defined by the deformed volume (Eulerian frame). By using the transformations, $e_{V'} = Je_V$, $\rho_{A'} = J\rho_A$, and $s_{V'} = Js_V$, we obtain the differential of e_V:

$$de_V = Tds_V - \frac{(P + f_v - \mu_{Av}\rho_A)}{J} dJ + \mu_{Av} d\rho_A, \tag{7}$$

where $f_V = e_V - Ts_V$ is the Helmhotlz free energy density (Eulerian frame). Finally, we rearrange the expression of de_V to arrive at the differential of s_V:

$$Tds_V = de_V + \frac{(P + f_V - \mu_{Av}\rho_A)}{J} dJ - \mu_{Av} d\rho_A. \tag{8}$$

3.2. Balance Equations

Here, we introduce the balance equations for mass, internal energy and entropy. We centralize the description of deformation by using a lattice velocity field, $\vec{v}_L$, which is the velocity of the imaginary network of indestructible sites called "markers" [5]. Consequently, the relationship between the lattice material time derivative and the partial derivative can be written as

$$\frac{d^L}{dt} = \frac{\partial}{\partial t} + \vec{v}_L \cdot \nabla. \tag{9}$$

Let us first consider the mass balance equation. In the absence of site creation and annihilation, the density of the material species i, where i can be A or v, changes through the equation:

$$\frac{\partial \rho_i}{\partial t} + \nabla \cdot \left(\rho_i \vec{v}_L + \vec{J}_i^L \right) = 0, \tag{10}$$

where $\vec{J}_i^L$ is the flux of the species i relative to the lattice. Using equation (9), we arrive at [5]

$$\frac{d^L \rho_i}{dt} + \rho_i \nabla \cdot \vec{v}_L + \nabla \cdot \vec{J}_i^L = 0. \tag{11}$$

For the energy balance equation, we first consider the fact that the total energy of the system comprises of the internal, kinetic, and potential energies; however, the latter two can be neglected if the system is stationary and is not influenced by an external field such as gravity. Therefore, the balance equation for the energy is [5]

$$\frac{\partial e_V}{\partial t} + \nabla \cdot \left(e_V \vec{v}_L + \vec{J}_E^L \right) = \dot{w}, \tag{12}$$

where $\dot{w}$ is the rate of mechanical work per unit volume and $\vec{J}_E^L$ is the flux of the internal energy relative to the lattice. In the absence of an applied field and the assumption of the mechanical equilibrium, we have $\dot{w} = \boldsymbol{\sigma} : \nabla \vec{v}_L$, where $\boldsymbol{\sigma}$ is the Cauchy stress tensor. Furthermore, for a system under hydrostatic pressure, $\bar{P}_{\text{ext}}$, the stress tensor becomes $\boldsymbol{\sigma} = -\bar{P}_{\text{ext}} I$, where I is the identity tensor. The mechanical work then reduces to $\dot{w} = -\bar{P}_{\text{ext}} \nabla \cdot \vec{v}_L$. Using the expression for $\dot{w}$ and equation (9), the energy balance equation becomes [5]

$$\frac{d^L e_V}{dt} + e_V \nabla \cdot \vec{v}_L + \nabla \cdot \vec{J}_E^L = -\bar{P}_{\text{ext}} \nabla \cdot \vec{v}_L. \tag{13}$$

Lastly, the entropy balance is postulated to be [5]

$$\frac{\partial s_V}{\partial t} + \nabla \cdot \left(s_V \vec{v}_L + \vec{J}_S^L \right) = \dot{s}, \tag{14}$$

where the entropy change is affected by the entropy flux relative to the lattice, $\vec{J}_S^L$, and entropy generation, $\dot{s}$, due to irreversibility. Using equation (9), we arrive at [5]

$$\frac{d^L s_V}{dt} + s_V \nabla \cdot \vec{v}_L + \nabla \cdot \vec{J}_S^L = \dot{s}. \tag{15}$$

3.3. Entropy Production

In this section, we derive the expression for the entropy production, $\dot{s}$, which will be used to identify the corresponding flux and force for the phenomenological relation in the next subsection. We consider the fundamental equation (8) derived earlier and replace the differential with the lattice material time derivative to obtain

$$T \frac{d^L s_V}{dt} = \frac{d^L e_V}{dt} + \frac{(P + f_V - \mu_{Av} \rho_A)}{J} \frac{d^L J}{dt} - \mu_{Av} \frac{d^L \rho_A}{dt}. \tag{16}$$

Using the balance equations (11) and (13), and the identity $d^L J/dt = J \nabla \cdot \vec{v}_L$, we arrive through lengthy but straightforward derivation

$$\frac{d^L s_V}{dt} + s_V \nabla \cdot \vec{v}_L + \nabla \cdot \vec{J}_S^L = -\vec{J}_Q^L \cdot \frac{\nabla T}{T^2} - \vec{J}_A^L \cdot \frac{\nabla \mu_{Av}}{T} + (P - \bar{P}_{\text{ext}}) \frac{\nabla \cdot \vec{v}_L}{T}, \tag{17}$$

where $\vec{J}_Q^L = T \vec{J}_S^L = \vec{J}_E^L - \mu_{Av} \vec{J}_A^L$ is the heat flux relative to the lattice. Compared the above equation with equation (15), we can identify the entropy production to be

$$T \dot{s} = -\vec{J}_Q^L \cdot \frac{\nabla T}{T} - \vec{J}_A^L \cdot \nabla \mu_{Av} + \nabla \cdot \vec{v}_L (P - \bar{P}_{\text{ext}}). \tag{18}$$

Following Ref. [11], the form of the above equation allows us to identify that one of the fluxes is $\nabla \cdot \vec{v}_L$ and the corresponding force is $P - \bar{P}_{\text{ext}}$.

3.4. Phenomenological Relations and Pressure-Controlled Dynamic Equation

Based on the classical irreversible thermodynamics framework, it is postulated that the forces and fluxes are related by the phenomenological (linear) law [12]:

$$J_\alpha = \sum_\beta L_{\alpha\beta} X_\beta, \tag{19}$$

where J_α represents thermodynamic fluxes, X_β represents thermodynamic forces and $L_{\alpha\beta}$ represents phenomenological coefficients. Using the assumption of symmetry properties of materials [5], the linear law reduces to

$$J_\alpha = L_{\alpha\alpha} X_\alpha. \tag{20}$$

Considering the flux and force identified in the previous subsection, we can write the phenomenological relation as

$$\nabla \cdot \vec{v}_L = L_P \cdot (P - \bar{P}_{\text{ext}}), \tag{21}$$

where L_P is the phenomenological coefficient. From the Reynolds transport theorem [6, 7], we can write the balance law as

$$\frac{dV}{dt} = \int_V (\nabla \cdot \vec{v}_L) d\vec{r}. \tag{22}$$

Substituting equation (21) into equation (22), we finally arrive at the pressure-controlled dynamic equation:

$$\frac{dV}{dt} = L_P \int_V (P - \bar{P}_{\text{ext}}) d\vec{r}. \tag{23}$$

4. Numerical Results

To highlight the difference between equation (3) (referred to as the KP equation) and equation (23) (referred to as the EP equation), we set $M = L_p V'$ (time scaling) and rearrange the KP equation to obtain $dJ/dt = L_P \left(\bar{P}_{\text{int}} - \bar{P}_{\text{ext}}\right) = L_P \Delta\bar{P}$, where $J = V/V'$ and V' is the volume at the initial state. For the EP equation, we assume that[4] $P = -f + \mu\rho$ and thus the EP equation becomes $dJ/dt = JL_P \left(\bar{P}_{\text{int}} - \bar{P}_{\text{ext}}\right) = JL_P \Delta\bar{P}$. From these alternative expressions of the KP and EP equations, one can see the extra J factor in the right-hand side of the EP equation. This extra factor causes the rate of deformation (dJ/dt) from the EP equation to be different from that of the KP equation. For tensile deformation from the initial state $(J(0) = 1)$, one can see that $J(t) > 1$ and the deformation rate from the EP equation will be higher than that from the KP equation. On the other hand, for compressive deformation, $J(t) < 1$ and the deformation rate from the EP equation will be lower than that from the KP equation. This behavior can also be shown from numerical calculations in figure 1. In figure 1(a), the system undergoes tensile deformation $(\Delta\bar{P} > 0)$ and the rate of increase in J from the EP equation is greater than that from the KP equation. In figure 1(b), the system is subjected to compressive deformation $(\Delta\bar{P} < 0)$ and the rate of decrease in J from the EP equation is now less than that from the KP equation.

[4] See Ref. [13] for a more rigorous definition of the pressure of the solid phase.

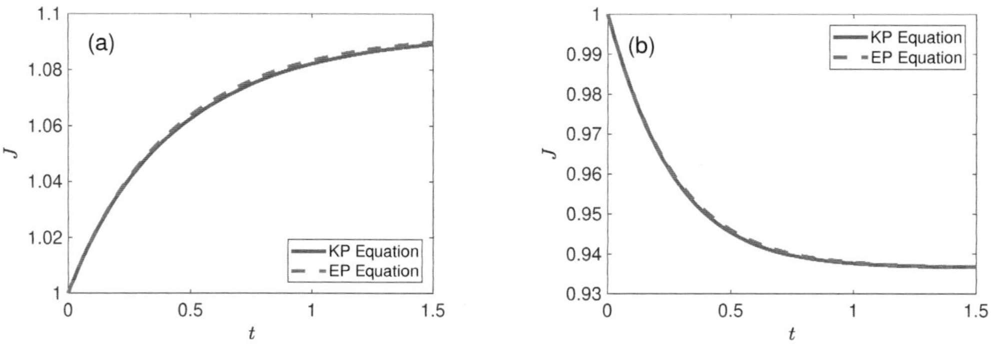

Figure 1. Comparison of time-dependent deformation behavior from equation (3) (KP equation) and equation (23) (EP equation). The system is subjected to constant external pressure that results in (a) tensile deformation ($\bar{P}_{\text{ext}} = 0.7\bar{P}_{\text{int}}|_{J=1}$) and (b) compressive deformation ($\bar{P}_{\text{ext}} = 1.3\bar{P}_{\text{int}}|_{J=1}$).

5. Summary

In this work, we developed the pressure-controlled dynamic equation for a single-component solid under applied hydrostatic pressure. The derivation is based on the established theories such as the thermodynamics of solids, kinematics of deformation, and non-equilibrium thermodynamics. The resulting equation (EP equation) exhibits non-equilibrium behavior that is different from that from the previously proposed pressure-controlled dynamics equation (KP equation). The deformation rate from the EP equation is higher (lower) than that from the KP equation for the tensile (compressive) deformation. The derivation presented in this work also provides a platform for further extension to describe a more complex system such as multi-component solid under non-hydrostatic stress.

Acknowledgments

This research was supported by the NSTDA-University-Industry Research Collaboration (NUI-RC) program under grant No. NUI-RC-M33-22-59-001D, King Mongkut's Institute of Technology Ladkrabang (KMITL) Research Fund under Grant No. KREF 015803, Thailand Research Fund (TRF) under Grant No. TRG5880008, and High Performance Computing Services from NECTEC (Thailand). The authors are indebted to Assoc.Prof.Wicharn Techidheera and Dr.Sasawat Mahabunphachai for their helpful discussion and support.

References

[1] Elder K R, Katakowski M, Haataja M and Grant M 2002 *Physical Review Letters* **88** 245701
[2] Emmerich H, Löwen H, Wittkowski R, Gruhn T, Tóth G I, Tegze G and Gránásy L 2012 *Advances in Physics* **61** 665–743
[3] Kocher G and Provatas N 2015 *Physical Review Letters* **114** 155501
[4] Larché F and Cahn J W 1973 *Acta Metallurgica* **21** 1051–1063
[5] Mishin Y, Warren J, Sekerka R and Boettinger W 2013 *Physical Review B* **88** 184303
[6] Malvern L E 1977 *Introduction to the Mechanics of a Continuous Medium* (Pearson)
[7] Rudnicki J W 2014 *Fundamentals of Continuum Mechanics* (John Wiley & Sons)
[8] De Groot S R and Mazur P 2013 *Non-Equilibrium Thermodynamics* (Courier Corporation)
[9] Wu K A and Karma A 2007 *Physical Review B* **76** 184107
[10] Voorhees P, Johnson W, Henry E and Frans S 2004 *Solid State Physics* **59**
[11] Balluffi R W, Allen S and Carter W C 2005 *Kinetics of Materials* (John Wiley & Sons)
[12] Lebon G, Jou D and Casas-Vázquez J 2008 *Understanding Non-Equilibrium Thermodynamics* (Springer)
[13] Chan V W L, Pisutha-Arnond N and Thornton K 2017 *Computational Materials Science* **135** 205–213

IOP Conf. Series: Materials Science and Engineering **361** (2018) 012007 doi:10.1088/1757-899X/361/1/012007

Comparison between some materials composites in the shock wave attenuation and anti-detonation property

Yiqing Fan, Jiong Wang

Nanjing University of Science & Technology in Nanjing, China

E-mail: wjiongz@njust.edu.cn

Abstract. Comparing the anti-detonation property and shock wave attenuation law of the single material structures, composite material structures and sandwich material structures are simulated by AUTODYN. The simulation on the structures with different combination show that, the aramid fiber material as base material in touched with the explosives, can greatly reduce the shock wave in identification board. For different material structures of the attenuation rule of shock wave, numerical simulation results show that the sandwich structure of the shock wave attenuation ability is more effectively than a single material structure. Based on the anti-detonation performance in different arrangement of the material structure, anti-detonation structure optimization design scheme is given according to the simulation results.

1. Introduction

With the wide application of insensitive high energy explosives, the powerful shock wave generated by the explosives has become an important means of killing enemy personnel in the battlefield environment. Therefore, it is of great significance to prevent or weaken the effect of shock wave on human body. In order to prevent the damage caused by impact, the protective layer absorbs most of the impact energy, so that the protected material is not damaged. It is very concerned to know the failure mode and strength of structure under load condition. The method of structural test is time-consuming and costly. With the development of computer science, the use of computer to simulate and predict the destruction process of structure, received a lot of scholars attention. Wang et al. [1] use LS-DYNA finite element software to simulate the dynamic response process of polyurea elastomer composite sandwich structure. Khurana et al. [2] supported theoretical calculation about some shock wave parameters by Autodyne 2D hydro-code simulation. Lan et al. [3] conducted a two-dimensional Lagrangian computational fluid dynamics program to simulate detonation wave propagation in PMMA. Kim et al. [4] defined a multi-material hydrodynamic simulation for LSCT to provide the detonation characteristics of high explosives in contact with a PMMA gap. Qin et al. [5] use LS-DYNA software to simulate the shock wave effect to evaluate flashing detonation grenade non-lethal effect. In this paper, the protective material with relative excellent detonation performance is obtained by designing different material composite structure forms. In this study, different combinations of materials were simulated.

2. Numerical Simulation Research on Shock Wave Attenuation Properties in single material

2.1. Computation model

Content from this work may be used under the terms of the Creative Commons Attribution 3.0 licence. Any further distribution of this work must maintain attribution to the author(s) and the title of the work, journal citation and DOI.
Published under licence by IOP Publishing Ltd

IOP Conf. Series: Materials Science and Engineering **361** (2018) 012007 doi:10.1088/1757-899X/361/1/012007

The dynamics element program AUTODYN is employed to simulate the explosive detonation attenuation rule of shock wave in different materials.

Modeling according to figure 1, explosive using RDX which selects the JWL [6] state equation describe the process of detonation product expansion process. Identification board materials choose several common protective materials such as steel plate, aluminum plate, aramid fiber composite material, concrete description. Due to the model is a symmetric structure, setting up a two-dimensional axial symmetric model. Array monitoring points are set in identification board axis. Center point initiation is chosen as the mode of detonation. Explosives and identification board using Lagrange solver.

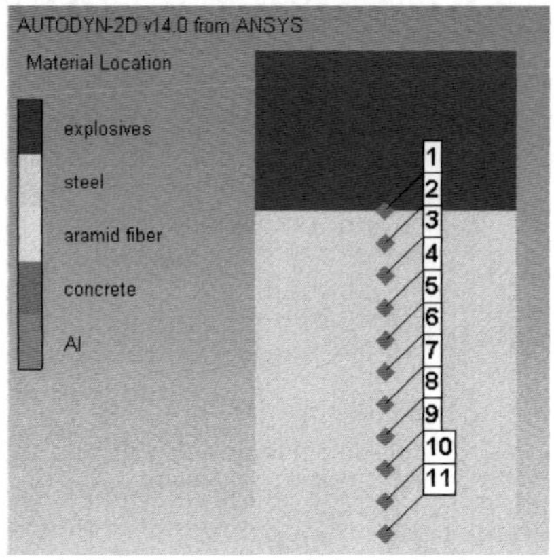

Figure 1. Simulation model of single material

2.2. Simulation results

Using the simulation method of section 2.1, the pressure distribution at different location of different material are shown in table 1.We can clearly observe from table 1 to get the information of the shock wave pressure of different material at different gauges.

Table 1. Simulation result of different gauges in single material structures

Gauge / Pressure (KPa) / Material	1	2	3	4	5	6	7	8	9	10	11
steel	7.21e7	3.21e7	1.76e7	1.03e7	6.77e6	4.55e6	3.92e6	3.54e6	3.13e6	2.35e6	2.51e4
Aramid fiber	4.72e6	2.58e5	1.23e5	5.24e4	6.55e3	3.95e3	5.06e3	4.35e2	1.13e2	19.294	3.9e-2
Concrete	3.64e7	1.25e7	3.71e6	2.4e6	1.93e6	1.1e6	5.78e5	3.42e5	1.98e5	1.22e5	5.73e3
Al	4.64e7	2.56e7	1.44e7	7.98e6	4.96e6	3.95e6	3.66e6	3.37e6	3.16e6	2.34e6	5.85e5

We can see from table 1 that the best shock wave attenuation materials are aramid fiber and concrete.

3. Numerical Simulation Research on performance of composite material structures

3.1. Computation model

We choose the two best performance materials in section 2.2 as the research object namely aramid fiber and concrete. Using the same modelling method in section 2.1, explosives and identification board size is changeless whereas identification board materials change from a single material into composite materials, recording detonation wave attenuation law shown as figure 2.

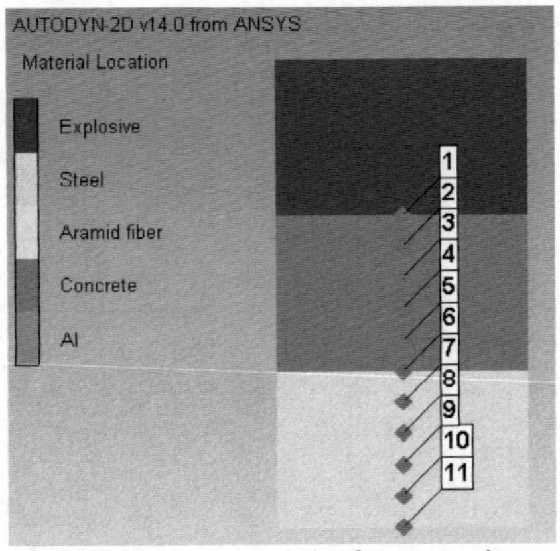

Figure 2. Simulation model of composite material

3.2. Simulation results

Gauges marked in the same position of identification board whose pressure changes as shown in table 2. Aramid fiber as base material in contact with the explosive has more strong attenuation ability than concrete as base material of contact with explosives.

Table 2. Simulation result of different gauges in composite material structures

Gauge / Pressure (KPa) / Composite	1	2	3	4	5	6	7	8	9	10	11
Concrete/ Aramid fiber	3.64e7	1.25e7	3.71e6	2.4e6	1.92e6	3.59e5	4.37e3	4.2e3	4.11e3	35.724	1.2e-1
Aramid fiber /Concrete	3.64e7	1.25e7	3.71e6	2.4e6	1.92e6	3.59e5	8.86e3	4.08e3	4.85e3	45.187	1.6e-1

4. Numerical Simulation Research on performance of sandwich material structure

4.1. Computation model

We still choose aramid fiber and concrete as the research object namely.Using the same simulation method in section 2.1, explosives and identification board size also remains the same. Change the identification board material distribution to sandwich structure and record shock wave attenuation law shown as figure 3.

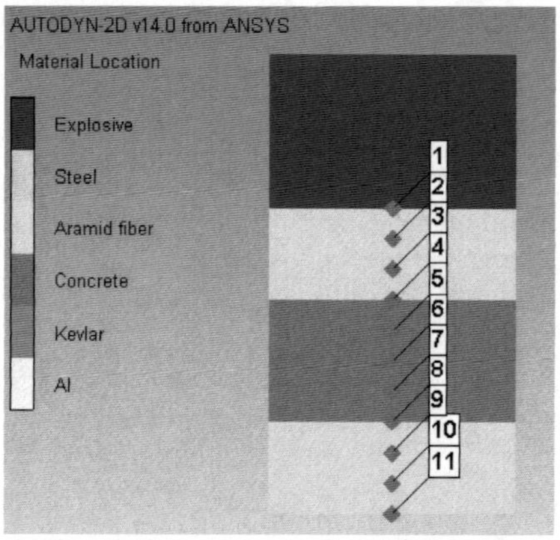

Figure 3. Simulation model of sandwich structure

4.2. *Simulation results*

Gauges in the sane marked position of the identification board whose pressure changes as table 3. Contrast table 1 and table 2, table 3 find the sandwich structure has strongest anti-detonation ability which detonation wave attenuation rate is about 88.98%.

Table 3. Simulation result of different gauges in sandwich material structures

Gauge / Pressure (KPa) / Sandwich	1	2	3	4	5	6	7	8	9	10	11
Aramid fiber /Concrete/ Aramid fiber3:4:3	4.72e6	1.84e5	1.07e5	1.29e5	1.71e5	7.37e4	3.62e4	93.651	5.7e-1	9.4e-2	1.4e-4

5. Theoretical analysis of simulation results

5.1. *Detonation wave propagation in composite material structure*

When detonation wave transmit form wave impedance $\rho_0 C_0$ medium to the wave impedance $\rho_1 C_1$ medium, direction of propagation is perpendicular to the interface, the intensity of detonation wave after interface transmission is P :

$$P = \frac{2}{1+\lambda} P_0, \quad \lambda = \frac{\rho_0 C_0}{\rho_1 C_1} \tag{1}$$

Because aramid fiber wave impedance is far less than the wave impedance of concrete [7-9], the shock waves attenuation rate from aramid fibers to the concrete is less than from aramid fiber into the concrete which is consistent with the results in table 2.

5.2. *Detonation wave propagation in sandwich material structure*

Using sandwich structure, in accordance with section 4.1 structure is analyzed. By the continuum mechanics, a sandwich structure attenuation of shock wave strength P' as follows:

$$P' = \left[\frac{2}{(1+\lambda)} \bullet \frac{2}{\left(1+\frac{1}{\lambda}\right)} \right] P_0 \tag{2}$$

Due to the formation of the sandwich structure, shock waves at the interface between different materials can be repeated reflection, further consumption of the shock wave energy. So the sandwich structure of shock wave attenuation ability than single structure.

6. Verification experiments

To confirm the validity of the proposed simulation results, a series of experiments are carried out by using eight types. Table 4 shows the experimental conditions. The simulation results traces precisely the experimental results in all cases.

Table 4. Comparison between experimental and simulation pressure

Material	Experiment Pressure(KPa)	Simulation pressure(KPa)
Explosive only	8e7	7.53e7
steel	2e4	2.51e4
Aramid fiber	0.04	3.9e-2
Concrete	5800	5.73e3
Al	5.9e5	5.85e5
Concrete/Aramid fiber	0.15	1.2e-1
Aramid fiber /Concrete	0.18	1.6e-1
Aramid fiber /Concrete/ Aramid fiber3:4:3	1.5e-4	1.4e-4

7. Summary

There are many ways to design anti-detonation material structure. The different simulation progress show single material structure can be chosen as the best two performance shock wave attenuation material ,concrete as the base material in touch with the explosive has better performance than aramid fiber and sandwich material structure has the best performance compared with single material structure and composite structure.

References

[1] Wang X, He J, Zu X, Wang D,Zhen J,Chen Q,Hu W and Zhai W 2017 Antidetonation Properties on composite sandwich structure with polyurea elastomer *Eng. Plast. Appl.*45 63
[2] Khurana R, Gautam P C, Rai R, Kumar A, Sharma A C and Singh M 2012 Studies on Shock Attenuation in Plastic Materials and Applications in Detonation Wave Shaping *J Phys Conf* 377 12051
[3] Lan W and Hu X 2012 Simulation Of Attenuation Regularity Of Detonation Wave In Pmma *AIP* 1426 331
[4] Kim B, Park J and Yoh J J 2016 Analysis on shock attenuation in gap test configuration for characterizing energetic materials *J Appl Phys* 119 145902
[5] Qin H and Guo S 2014 Simulation of the shock wave effect of flashing detonation grenade *J Liao Tech U* 33 1497
[6] Kato H, Kaga N, Takizuka M, Hamashima H and Ito S 2004 Research on the JWL parameters of several kinds of explosives *Mater Sci Forum* 465-466 271
[7] Shi S Q, Liu R H and Min W 2008 Shock wave reduction behavior of a new compound structure composed of a foam aluminumlayer between two steel plates *J Vib & Shock* 27 143
[8] Yang X 2007 Study on anti-detonation properties of armor plate/kevlar composites *New Chem Mater* 35 36
[9] Zhang L and Hu S 2006 A novel experimental technique to determine the spalling strength of concretes *Explo Shock W* 26 537

IOP Conf. Series: Materials Science and Engineering **361** (2018) 012008 doi:10.1088/1757-899X/361/1/012008

Effects of copper content in Sn-based solder on the intermetallic phase formation and growth during soldering

N Mookam, P Tunthawiroon and K Kanlayasiri

Department of Industrial Engineering, Faculty of Engineering,
King Mongkut's Institute of Technology Ladkrabang, Bangkok 10520, Thailand

Email: niwat_ie@hotmail.com

Abstract. This research has studied the effects of copper contents in the Sn-based solder (i.e. Sn-0.7Cu, Sn-1.0Cu, and Sn-3.0Cu) on the intermetallic phase formation and the growth during soldering. In the experiments, reflow soldering was performed at 350 °C with the Cu substrate by using three solders for various soldering times of 10, 20, 40, 60, 120, 240 and 480 s. XRD data have shown that the η-Cu_6Sn_5 and ε-Cu_3Sn phases with hexagonal crystal structure were present between the solder and the substrate. The presence of ε-Cu_3Sn phase was found only when the substrate was soldered for at least 240 s. The Cu content in solder had significant effects on the intermetallic growth with long term soldering time. In addition, the diffusion coefficient was correlated to the Cu content in the solder.

1. Introduction

Copper (Cu) is widely used as an electrical conductor in circuit boards, due to its high electrical and thermal conductivity [1]. Solder alloy is used to connect the electronic devices onto the Cu substrate. Among tin (Sn)-based solder systems, the Sn–Cu solder is considered as the most promising candidate to replace the Sn-Pb solder compared to other solders because of their low melting temperature and favorable properties [2]. In soldering processes, where a solder is in contact with a Cu substrate, there are phenomena involving the development of a nucleus and the subsequent formation and the growth of intermetallic compound (IMC). Hence, the reaction between the Cu and Sn components in an equilibrium state induces the IMC beginning to form at the interface of the solder and substrate [3, 4].

It is well known that the Cu-Sn soldering system has two interfacial phases which are Cu_6Sn_5 and Cu_3Sn phases [5]. At lower temperatures, the crystal phase of Cu_6Sn_5 is monoclinic. However, this η'-Cu_6Sn_5 phase can transform to hexagonal η-Cu_6Sn_5 phase with higher temperature [6]. The η-Cu_6Sn_5 phase transforms from the hexagonal to monoclinic during the cooling procedure. In terms of ε-Cu_3Sn phase, the crystal structure is a hexagonal sphere packing with A3 structure type [7]. Besides, there are phase transitions from A3 structure to the D019 structure during quenching process from high temperature. Furthermore, the ε-Cu_3Sn phase has an orthorhombic structure which develops during the solidification process of the Cu–Sn alloy. As the ε-Cu_3Sn phase is often described as an orthorhombic crystal system due to orthorhombic unit cell, and thus constitutes to the hexagonal structure [7, 8].

The microstructure and reaction interface changes were influenced by increasing the ratio of Cu to solder [9]. Higher Cu compositions in the solder assist the formation and increase the growth of Cu_6Sn_5 instantly [10, 11]. Typically, IMCs are less ductile which is related to their low thermal and low electrical conductivity. The low ductility (i.e. brittleness) of the IMCs contributes to the fatigue

Content from this work may be used under the terms of the Creative Commons Attribution 3.0 licence. Any further distribution of this work must maintain attribution to the author(s) and the title of the work, journal citation and DOI.
Published under licence by IOP Publishing Ltd

lifetime reduction of solder joints. In other words, the presence the growth of IMCs has an impact on the reliability of solder joints [12, 13]. Nonetheless, the formation and growth of IMC at the interface are influential of various factors, including the solder alloy and reflow conditions [14]. Despite several studies on the concentration of the Cu in the solder, the effects of a higher concentration of Cu in the solder have not been shown, especially with 3.0 wt.% of Cu in the Sn-based solder. The aim of this study is to investigate the effects of Cu contents in the Sn-based solder on the IMC formation and the growth during reflow soldering.

2. Experiment

2.1. Materials and methods

The experiments were carried out using Sn-based solder with 0.7, 1.0 and 3.0 wt% of Cu, termed Sn-0.7Cu, Sn-1.0Cu and Sn-3.0Cu respectively. The solder bar was casted to have a diameter of 6.5 mm and a thickness of 1.24 mm (JIS Z3198-3). The Cu substrates with purity of 99.99% (JIS H3100) were 0.2 mm thick and 25×30 mm. Prior to reflow soldering, the substrate was cleaned using dilute HCI acid and flux (RC-15SH RMA (15%)) was droplets onto the substrate. For the soldering process, the substrate was placed onto a hot plate followed by the solder. Next, the temperature was increased to 350 °C for 10, 20, 40, 60, 120, 240 and 480 s respectively. After soldering, samples were mounted and polished respectively. The polished surface was etched with a solution containing of 93% ethanol, 5% nitric acid and 2% HCI. Samples were then analysed with the following techniques. Firstly, scanning electron microscopy (SEM, JEOL; JSM-5800LV) was used to verify the microstructure and intermetallic phases. Secondly, an energy dispersive spectroscopy (EDS, OXFORD INSTRUMENTS, X-Max) was used to determine the chemical compositions of the IMCs. Finally, X-ray diffraction (XRD, Bruker, D8-Discover) was carried out to analyze the crystal and lattice structure parameters of IMCs.

3. Results and Discussion

3.1. Intermetallic phase formed during soldering

During reflow soldering, the Cu substrate is in contact with solder bringing about the prediction of the IMCs generated from the Sn-Cu binary phase diagram. As a result, two intermetallic layers which are Cu_6Sn_5 and Cu_3Sn phase are clearly visible at the temperature of 350 °C in the soldering process while other Cu-Sn IMCs, such as $Cu_{10}Sn_3$ and $Cu_{41}Sn_{11}$, are found at the higher temperature [6]. The IMCs of Cu_6Sn_5 and Cu_3Sn were introduced by the reaction between the Sn atoms from the solder and Cu atoms from the substrate [15, 16] which can be explained by Equations (1)-(2):

$$6Cu + 5Sn \rightarrow Cu_6Sn_5 \quad (1)$$
$$3Cu + Sn \rightarrow Cu_3Sn \quad (2)$$

The diffusion of atomic species, Sn and Cu passing through the interface results in the formation of the first Cu_6Sn_5 layer followed by the Cu_3Sn layer adjacent to the Cu substrate [17]. Meanwhile, the Cu_3Sn could react with Sn atoms to form Cu_6Sn_5 [18]. According to Equation (3), Sn is released in liquid state. However, it was reported that Cu_6Sn_5, which is a kinetically favorable phase, is unstable with Cu, but Cu_3Sn, as a thermodynamically steady phase, is stable with Cu. Thus, if there is free Cu in the system, Cu_6Sn_5 tends to decompose. This can occur by two ways, either by adding Cu or by subtracting Sn from Cu_6Sn_5. The chemical reactions converting Cu_6Sn_5 to Cu_3Sn are shown in Equations (4)-(5) [18, 19].

$$2Cu_3Sn + 3Sn \rightarrow Cu_6Sn_5 \quad (3)$$

$$Cu_6Sn_5 + 9Cu \rightarrow 5Cu_3Sn \quad (4)$$
$$Cu_6Sn_5 \rightarrow 5Cu_3Sn + 3Sn \quad (5)$$

Nevertheless, addition of Cu by Equation (4) results in a Cu_3Sn layer between the Cu_6Sn_5 layer and the Cu substrate where the Cu atom refers to the dissolved substrate [20]. Thus, η-Cu_6Sn_5 forms first and ε-Cu_3Sn forms later at the η-Cu_6Sn_5/Cu interface. In the aspect of the depletion of Sn from η-Cu_6Sn_5 in Equation (5), the Sn tends to diffuse to the side of Cu to form ε-Cu_3Sn. However, there must be a sink for Sn or the presence of a nearby Cu to attract Sn. From previous research, many phenomena occurred which were related to the ε-Cu_3Sn and associated with the size of the IMC. Thus, we shall discuss and conclude with the thickness of IMC details.

The SEM observation was carried out to study the formation of IMCs at various soldering times and solder compositions as shown in Figure 1. In the experiment, similar IMC formation was found between solder and Cu substrate. Two types of IMCs (η-Cu_6Sn_5 and ε-Cu_3Sn) were found at the interface. However, ε-Cu_3Sn was found only when the Cu substrate has been soldered for longer soldering times (240 s and 480 s), while η-Cu_6Sn_5 was found at other soldering conditions (10 s-120 s).

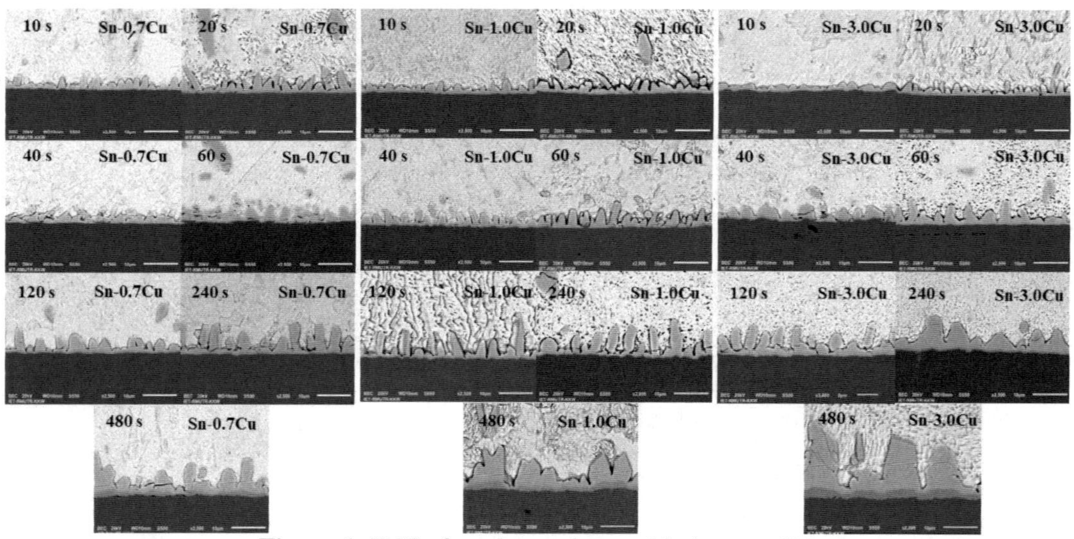

Figure 1. IMCs found at various soldering conditions.

The chemical composition of both IMCs was determined via EDS analysis as illustrated in Figure 2. The findings indicated that Cu contents of the *η*-Cu_6Sn_5 phase range from 51.11 at.% to 52.77 at.%, the region of ε-Cu_3Sn an elemental content of 70.74 at.% to 73.38 at.% Cu and respective Sn is found. The confirmation of phase formation was determined via XRD analysis as illustrated in Figure 3. From the XRD pattern of the specimen, the Cu phase is α-Cu which consists of a face-centered cubic structure, and the solder is β-Sn which adopts a body centered tetragonal structure. The crystal structure of both η-Cu_6Sn_5 and ε-Cu_3Sn is hexagonal. The *a*, *b* and *c* lattice parameters of *η*-Cu_6Sn_5 were 4.200, 4.200 and 5.090 angstrom. The η-Cu_6Sn_5 phase conforms to a B81 structure and a space group of P63/mmc. The lattice parameters of ε-Cu_3Sn displays the *a* and *b* were 2.749 and 4.322 angstrom respectively. However, η-Cu_6Sn_5 did not transform to η'-Cu_6Sn_5 during the cooling period of the reflow.

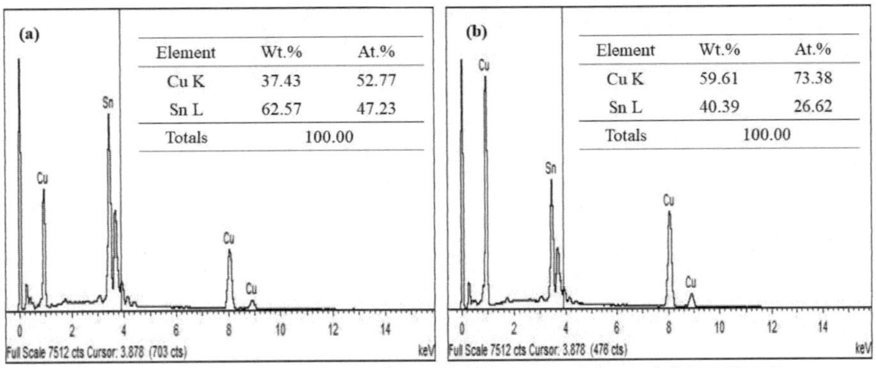

Figure 2. EDS spectrum for Sn-3.0Cu: (a) η-Cu_6Sn_5 and (b) ε-Cu_3Sn.

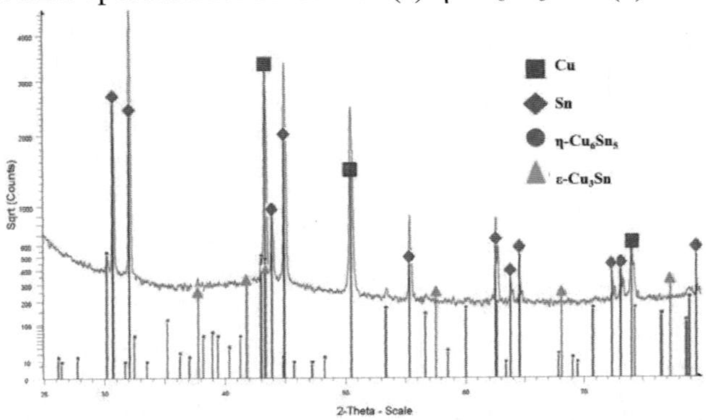

Figure 3. XRD pattern of specimen soldered at 480 s for Sn-3.0Cu.

3.2. Thickness of intermetallic compound

The thickness of IMCs at different soldering conditions was also examined, the results are shown in Figure 4. The findings indicated that with a soldering time of 10 s there is no difference in the thickness of η-Cu_6Sn_5 layer, Figure 4 (a). Due to the short soldering time, the supply of Cu and Sn atoms to their corresponding interfaces was composed of two parts, released from substrate or solder, and the decomposition of IMC phase. Following soldering at 20-40 s, the thickness of the IMC of both solders between Sn-0.7Cu and Sn-1.0Cu were not different, since both solders have slightly different Cu compositions. In contrast, the thickness behavior of Sn-3.0Cu solder was different, the thickness of η-Cu_6Sn_5 at the interface was significantly higher than those in the Sn-0.7Cu/Cu and Sn-1.0Cu/Cu. As the soldering time extended from 60 s to 120 s, the IMC thickness increased as a function of reflow time and Cu content. The Cu atoms can also diffuse from the substrate into solder through the intermetallic layer, and then react with Sn atoms to form huge Cu_6Sn_5 precipitates within the solder matrix. However, Cu atoms from substrate and solder not only reacted with Sn atoms diffused to promote the continuous growth of η-Cu_6Sn_5 layer [21]. Nevertheless, evidence of the transition from η-Cu_6Sn_5 to ε-Cu_3Sn according to Equation (4) was not found.

At a soldering time of 240 s, η-Cu_6Sn_5 tends to decompose, indicated by a reduced thickness of the η-Cu_6Sn_5. The Sn could diffuse through the η-Cu_6Sn_5 to react with Cu, it formed the ε-Cu_3Sn phase. Moreover, η-Cu_6Sn_5 tended to decompose due to subtracting Sn from η-Cu_6Sn_5, as illustrated by Equation (5). Thus, experiments have found that η-Cu_6Sn_5 forms firstly and is followed by ε-Cu_3Sn at the η-Cu_6Sn_5/Cu interface after a certain period of soldering. Figure 4 (b) illustrate the thicknesses of ε-Cu_3Sn layers when soldered at 240 and 480 s. The ε-Cu_3Sn thickness tended to increase with the expansion of the Cu content and soldering time. In addition, Cu is the dominant diffusing species in the ε-Cu_3Sn layer. Therefore, Cu atoms diffuse through the ε-Cu_3Sn layer and react with the existing η-Cu_6Sn_5 to form the ε-Cu_3Sn growth, as expressed in Equation (4). Thus, the formation and growth of ε-Cu_3Sn agrees with Equations (5) and (4) respectively. In addition, the ε-Cu_3Sn will grow at the expense of the η-Cu_6Sn_5 phase. The total thickness of the intermetallic layer is the sum of the thicknesses of ε-Cu_3Sn and η-Cu_6Sn_5 layers as shown in Figure 4 (c). The enhancement in the thickness of one phase proportionally reduces the thickness of the other phases in order; consequently, the growths of η-Cu_6Sn_5 and ε-Cu_3Sn are interrelated.

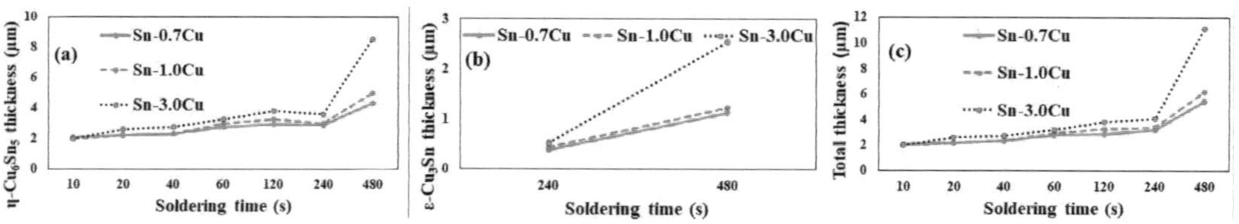

Figure 4. Thickness of IMCs: (a) η-Cu_6Sn_5, (b) ε-Cu_3Sn, and (c) η-Cu_6Sn5 + ε-Cu_3Sn.

3.3. Calculation of diffusion coefficients

The growth of the intermetallic compounds during soldering condition can be expressed by the Equation:

$$Y = Y_o + \sqrt{Dt} \tag{6}$$

where Y_o is the thickness of the intermetallic compound at t = 0, D is the diffusion coefficient. The value of $D^{1/2}$ could be determined from the slope of plot between the thickness of the intermetallic layer (Y) and the square root of soldering time ($t^{1/2}$).

Table 1 presents the comparison of the effective diffusion coefficients of the IMCs for Sn-0.7Cu/Cu, Sn-1.0Cu/Cu and Sn-3.0Cu/Cu solder joints. It was observed that the diffusion coefficient was correlated to the Cu content in the solder. Since the diffusion coefficients for the ε-Cu_3Sn of this study were higher than the η-Cu_6Sn_5. Due to the formation and growth of ε-Cu_3Sn phase is the sum of the product of Equations (4) and (5), and ε-Cu_3Sn phase increases in thickness at the expense of η-Cu_6Sn_5 phase.

Table 1. Effective diffusion coefficients of intermetallic compound.

Solders	Cu_6Sn_5 (um^2/s)	Cu_3Sn ($\mu m^2/s$)
Sn-0.7Cu	0.15340	0.370
Sn-1.0Cu	0.20410	0.395
Sn-3.0Cu	0.40785	1.010

The finding indicates that the intermetallic phase is fast for the formation and growth with increment of Cu content in solder. The higher Cu content of solder promotes Cu diffusion flux in bulk solder, and promotes nucleation of the interface. Nucleation of the interface can come from both new combination and pre-existing clusters to form IMC [22]. Moreover, the η-Cu_6Sn_5 growth at the solder/η-Cu_6Sn_5 interface was promoted due to the larger amount and size of Cu-Sn clusters. In summary, this study demonstrates the effects of differences in Cu content on the growth of IMC, which can be used as a guide for the selection of solder alloys in Sn-based solder.

4. Conclusions

This research studied the effects of Cu content in the Sn-base solder on the formation and growth of IMCs. XRD results showed η-Cu_6Sn_5 phase forms first followed by ε-Cu_3Sn at the η-Cu_6Sn_5/Cu interface. As soldering time increased from 240 s to 480 s, the intermetallic phase of the joint transformed from a single η-Cu_6Sn_5 phase to coexisting η-Cu_6Sn_5 and ε-Cu_3Sn phase. However, the Cu content in the solder had significant effects on the IMC growth with the long-term soldering time. The diffusion coefficient was correlated to the Cu content in the solder. The diffusion coefficient of the solder joints is high when the Cu content in the solder increases.

References

[1] So W W, Choe S, Chuang R and Lee C C. 2000 An effective diffusion barrier metallization process on copper *Thin Solid Films* vol 376 pp 164-169.

[2] Alam S N, Mishra P and Kumar R. 2015 Effect of Ag on Sn–Cu and Sn–Zn lead free solders *Materials Science-Poland* vol 33 pp 317-330.

[3] Sharif A, Chan Y C and Islam R A, 2004 Effect of volume in interfacial reaction between eutectic Sn–Pb solder and Cu metallization in microelectronic packaging *Materials Science and Engineering* vol 106 pp 120-125.

[4] Park M S, and Arro´yave R 2012 Concurrent nucleation, formation and growth of two intermetallic compounds (Cu_6Sn_5 and Cu_3Sn) during the early stages of lead-free soldering *Acta Materialia* vol 60 pp 923-934.

[5] Ramli, M I I, Saud N, Salleh M M A A and Derman M N and Said R M 2016 Effect of TiO_2 additions on Sn-0.7Cu-0.05Ni lead-free composite solder *Microelectronics Reliability* vol 65 pp 255-264.

[6] Okamoto H, Schlesinger M E and Mueller E M. Alloy Phase Diagrams. *ASM Handbook* Volume 3.

[7] Sang X, Du K and Ye H 2009 An ordered structure of Cu_3Sn in Cu–Sn alloy investigated by transmission electron microscopy *Journal of Alloys and Compounds* vol 469 129-136.

[8] Dibritz S, Hauffe W and Kurt R 1997 Investigation of the Three-Dimensional Microstructure of Cu-Sn(Pb) Diffusion Zones by Means of Ion Beam Sputtering, Scanning Electron Microscopy and Lattice Source Interferences *Mikrochim Acta* vol 125 pp 3-12.

[9] Hwang C-W and Suganuma K 2004 Effect of Cu Addition to Sn-Ag Lead-Free Solder on Interfacial Stability with Fe-42Ni *Materials Transactions* vol 45 pp 714-720.

[10] Rizvi M J, Lu H and Bailey C 2009 Moldeling the diffusion of solid copper into liquid solder alloys. *Thin Solid Flims* vol 517 1686-1689.

[11] Yuan Y, LI D, Guan Y, Seifert H J and Moelans N 2016 Investigation of the diffusion behavior in Sn-xAg-yCu/Cu solid state diffusion couples *Journal of Alloys and Compounds* vol 686 pp 794-802.

[12] Yang Y, Li Y, Lu H, Yu C and Chen J, 2013 Interdiffusion at the interface between Sn-based solders and Cu substrate *Microelectronics Reliability* vol 53 pp 327-333.

[13] Li Q and Chan Y C 2013 Growth kinetics of the Cu_3Sn phase and void formation of sub-micrometre solder layers in Sn–Cu binary and Cu–Sn–Cu sandwich structures *Journal of Alloys and Compounds* vol 567 pp 47-53.

[14] Li J F, Agyakwa P A and Johnson C M 2012 Effect of trace Al on growth rates of intermetallic compound layers between Sn-based solders and Cu substrate *Journal of Alloys and Compounds* vol 545 pp 70-79.

[15] Hu X, Li Y and Min Z 2014 Interfacial reaction and IMC growth between Bi-containing Sn0.7Cu solders and Cu substrate during soldering and aging *Journal of Alloys and Compounds* vol 582 pp 341-347.

[16] Yao P, Li X, Liang X, Yu B, Jin F and Li Y. 2017 A study on interfacial phase evolution during Cu/Sn/Cu soldering with a micro interconnected height *Materials Characterization* vol 131 pp 49-63.

[17] Flanders D R, Jacobs E G and Pinzzotto R F 1997 Activation Energies of Intermetallic Growth of Sn-Ag Eutectic Solder on Copper Substrates *Journal of Electronic Materials* vol 26 pp 833-887.

[18] Yu C, Lu H and Li S 2008 Effect of Zn addition on the formation and growth of intermetallic compound at Sn–3.5 wt% Ag/Cu interface *Journal of Alloys and Compounds* vol 460 pp 594-598.

[19] Wang Y, Chu D T and Tu K-N. 2016 Porous Cu_3Sn formation in Cu-Sn IMC-based micro-joints *IEEE 66th Electronic Components and Technology Conference* pp 439-446.

[20] Tang W-M, He A-Q, Liu Q and IVEY D G 2010 Solid state interfacial reactions in

electrodeposited Cu/Sn couples. Trans *Nonferrous Met. Soc. China* vol 20 pp 90-96.

[21] Wang F, Huang Y, Zhang Z and Yan C 2017 Interfacial Reaction and Mechanical Properties of Sn-Bi Solder joints. Materials, 10, 920; doi:10.3390/ma10080920.

[22] Haitao M, Haoran M, Anil K, Shengyan S, Yunpeng W, Jun C, Mingliang H and Ning Z Effect of initial Cu concentration on the IMC size and grain aspect ratio in Sn–xCu solders during multiple reflows *Journal of Materials Science: Materials in Electronics* DOI 10.1007/s10854-017-7952-9.

Acknowledgements

The authors would like to express sincere gratitude to King Mongkut's Institute of Technology Ladkrabang (KMITL)'s Faculty of Engineering for the financial support.

IOP Conf. Series: Materials Science and Engineering **361** (2018) 012009 doi:10.1088/1757-899X/361/1/012009

Investigation on Viscoelastic-Creep Behavior of the Phase-Field Crystal Method

J Em-Udom and N Pisutha-Arnond[1,2]

Department of Industrial Engineering, Faculty of Engineering, King Mongkut's Institute of Technology Ladkrabang, Bangkok 10520, Thailand

E-mail: [1]nirand.pi@kmitl.ac.th

Abstract. The phase-field crystal (PFC) method is a promising computational model with atomistic resolution and diffusive time-scale. In this study, we investigated the viscoelastic-creep behavior exhibited by the PFC model. We considered a one-dimensional crystal subjected to step changes in pressure and studied the time-dependent strain response from the system. The parametric study shows that the PFC model predicts the materials with higher density to exhibit higher stiffness and lower damping capacity while an increase in temperature results in lower stiffness and damping capacity. These predictions agree with experimental observations and show promising capability of the PFC method to model viscoelastic phenomena.

1. Introduction

Due to rapid increase in computing performance and theoretical development, computational methods are becoming powerful tools in facilitating materials advancement. However, many problems in computational materials science involve length and time scales of many orders of magnitude. This large span of length and time scales is a challenge for atomistic models, such as molecular dynamics (MD) and density functional theory (DFT), in that these models are typically applicable in prohibitively-small length and time regimes. On the other hand, continuum models, such as the phase field (PF) models, naturally reside in phenomenologically-relevant length and time scales, but often suffer from lack of atomistic resolution and self-consistency. In the past decade, the phase-field crystal (PFC) method [1] emerged as a robust tool to address these issues. The appeal of the PFC method lies in its atomistic resolution and diffusive time scale, which give the model advantages over the PF method in terms of accuracy and self-consistency and over the MD and DFT in terms of computational expenses. Due to these advantages, the PFC method has been used to predict various materials properties and model several materials phenomena (see Ref. [2] for comprehensive review in this topic).

Despite various applications of the PFC method, especially in deformation-related phenomena, the viscoelastic behavior exhibited by the model has not been extensively studied. Viscoelasticity is a combination of elasticity and viscosity, and is an important characteristic in vibration-damping materials, noise-controlled applications, and microelectromechanical system (MEMS). Therefore, in this work, we investigated the viscoelastic behavior exhibited by the PFC model. In particular, we considered the viscoelastic-creep behavior of a one-dimensional crystalline solid subjected to step changes in pressure, and analyzed the time-dependent strain

[2] Corresponding author

Content from this work may be used under the terms of the Creative Commons Attribution 3.0 licence. Any further distribution of this work must maintain attribution to the author(s) and the title of the work, journal citation and DOI.
Published under licence by IOP Publishing Ltd

IOP Conf. Series: Materials Science and Engineering **361** (2018) 012009 doi:10.1088/1757-899X/361/1/012009

response of the system. From the parametric study, we found that the PFC method predicts higher-density materials to have higher stiffness and damping capacity. Also, an increase in temperature results in lower stiffness and lower damping capacity. These predictions agree with experimental observations and show promising capability of the PFC method to model viscoelastic phenomena. Nevertheless, further investigation is needed in order to better understand the viscoelstic behavior from the PFC model; future improvements include extension to three-dimensional structures, more extensive parametric study, and investigation into other viscoelastic phenomena.

The paper is organized as follows. In section 2, we review the PFC method and the pressure-controlled dynamic equation recently proposed by Em-Udom and Pisutha-Arnond [3]. In section 3, we describe the simulation and analysis procedures. The simulation results and discussion are presented in section 4. Lastly, we provide a summary of our work in section 5.

2. Phase-Field Crystal Method

In the PFC method, the two central ingredients are the energy expression and evolution equations. For the energy expression, we consider the Helmholtz free energy, F, which can be written in terms of dimensionless quantities as [4]

$$F = \int_V f_v d\vec{r} \equiv \int_V \left(\frac{\rho^4}{4} + \frac{\rho}{2} \left[-\epsilon + \left(1 + \nabla^2\right)^2 \right] \rho \right) d\vec{r}, \tag{1}$$

where f_v is the free energy density, $\rho \equiv \rho(\vec{r})$ is the atomic number density field, ϵ is the model parameter related to temperature, and V is the volume. This free energy is minimized by a constant density profile (representing a liquid phase) and a periodic density profile (representing a crystalline phase). The latter profile can be written in terms of a Fourier expansion of the form

$$\rho(\vec{r}, t) = \bar{\rho}(t) + \sum_j A_j(t) e^{i\vec{G}_j \cdot \vec{r}} + \text{c.c.}, \tag{2}$$

where $\vec{G}_j$ is the reciprocal lattice vector, A_j is the amplitude of the density wave corresponding to $\vec{G}_j$, $\bar{\rho}$ is the average number density, and c.c. denotes the complex conjugate. The second ingredient of the PFC method, which is the evolution equations, governs the time evolution of the state of the system. In this work, we employ two evolution equations. The first is the Cahn-Hilliard type equation that evolves $\rho(\vec{r}, t)$ through dissipative dynamics and with mass conservation [5]:

$$\frac{\partial \rho}{\partial t} = L_\mu \nabla^2 \mu \equiv L_\mu \nabla^2 \left(\frac{\delta F}{\delta \rho} \right), \tag{3}$$

where $\mu = \delta F / \delta \rho$ is the chemical potential, L_μ is the mass-diffusion coefficient. The second equation is the pressure-controlled dynamic equation recently proposed in Ref. [3] to simulate the deformation of a system under externally-applied pressure $\bar{P}_{\text{ext}}$:

$$\frac{dV}{dt} = L_P \int_V (P - \bar{P}_{\text{ext}}) d\vec{r}, \tag{4}$$

where $P = -f_v + \mu\rho$ is the internal pressure and L_P is the deformation coefficient.

3. Method

In this section, we describe the method to simulate, analyze and study the viscoelastic-creep behavior in the PFC model. To simulate the viscoelastic-creep behavior, we consider a one-dimensional crystalline phase with the density field $\rho(x, t) = \bar{\rho}(t) + \sum_j A_j(t) \cos(G_j x)$, where x

is the spatial coordinate. The system is initially at equilibrium and this state of the system is characterized by three state variables. First is the mass per unit-cell $N_0 = \bar{\rho}_0 L_0$, where L_0 is the initial unit-cell length and $\bar{\rho}_0 \equiv \bar{\rho}(0)$ is the initial average density. The second state variable is the average pressure $\bar{P}_0 = 1/L_0 \int_{L_0} P dx$, and the third is the temperature which is related to the model parameter ϵ. Then, the system is subjected to step changes in the average pressure quantified by $\Delta\bar{P} = \bar{P}_0 - \bar{P}_{\text{ext}}$. From equation (4), the nonzero $\Delta\bar{P}$ will lead to deformation characterized by the change in the unit-cell length $L(t)$ and the strain $e(t) = (L(t) - L_0)/L_0$. During the deformation, we enforce mass conservation by changing the average density according to $\bar{\rho}(t) = N_0/L(t) = \bar{\rho}_0 L_0/L(t)$, and we also allow the system to attain the equilibrium condition in all deformation states by evolving the density profile using equation (3) until a steady state is reached for each value of $L(t)$.

To analyze the viscoelastic-creep behavior, we fit $e(t)$ from the simulations to the solution of a standard linear solid model [6] which is typically used to describe viscoelastic behavior. In particular, we consider a spring-damper system subjected to uniform stress σ as shown in figure 1. The strain response from this system is governed by $de/dt + \frac{E_1 E_2}{\eta(E_1+E_2)} e = \frac{E_1}{\eta(E_1+E_2)} \sigma$, where E_1 and E_2 are the spring constants and η is the damping coefficient [6]. The solution to this ordinary differential equation is

$$e(t) = e_\infty \left[1 - \exp\left(-\frac{t}{\tau}\right)\right], \quad e_\infty = \frac{\sigma}{E_2}, \quad \tau = \eta\left(\frac{E_1 + E_2}{E_1 E_2}\right), \tag{5}$$

where e_∞ and τ are the strain value at $t = \infty$ and the relaxation time, respectively. When equation (5) is fitted to $e(t)$ from the PFC simulations, the values of e_∞ and τ, which characterize the viscoelastic-creep behavior, can be extracted for further study.

To study the viscoelastic-creep behavior exhibited by the PFC model, we investigate the dependence of e_∞ and τ on the PFC model parameters. In this parametric study, we use $\Delta\bar{P} = \pm 0.05\bar{P}_0$, $\epsilon = [0.1, 0.3]$, and $|\bar{\rho}_0| = [0.15, 0.5]$. The values of $\Delta\bar{P}$ are selected to yield tensile ($\Delta\bar{P} > 0$) and compressive ($\Delta\bar{P} < 0$) deformation. The ranges of ϵ and $|\bar{\rho}_0|$ are selected from the crystalline region in the phase diagram shown in Ref. [5]. Other fixed simulation and model parameters are $L_\mu = 1$, $L_P = 0.5$, $L_0 = 2\pi\sqrt{2}$, Δx (grid spacing) $= L_0/16$. The time steps for equations (3) and (4) are 0.01 and 0.5, respectively.

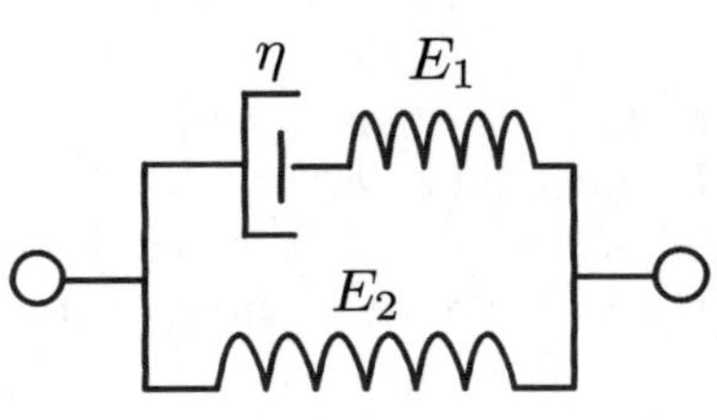

Figure 1. Schematic illustration of a spring-damper system.

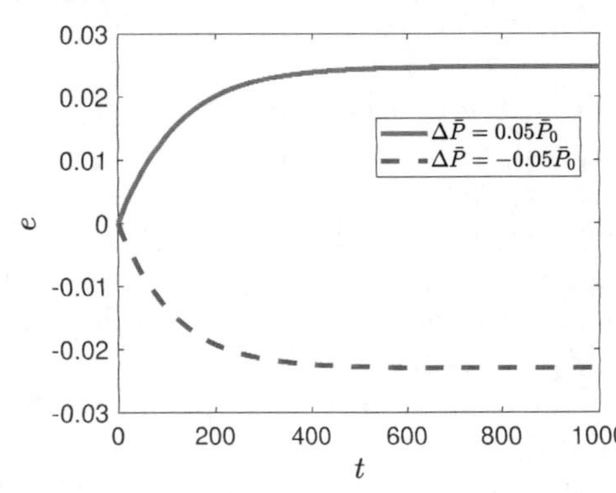

Figure 2. Plot of strain responses from the step changes in the pressure $\Delta\bar{P} = \pm 0.05\bar{P}_0$. In both simulations, $\epsilon = 0.3$ and $|\bar{\rho}_0| = 0.15$.

4. Results and Discussion

In figure 2, the viscoelastic-creep behavior from the PFC model is shown from the plot of the strain $e(t)$ with $\Delta\bar{P} = \pm 0.05\bar{P}_0$. Initially, the system is at $e = 0$. After the pressure changes, e increases or decreases exponentially until the convergence value is reached. The positive e indicates tensile deformation which results from the positive $\Delta\bar{P}$; on the other hand, the negative e implies compressive deformation which corresponds to the negative $\Delta\bar{P}$. This time-dependent strain response from a step change in pressure is indicative of the viscoelastic-creep behavior.

In figure 3, the dependence of e_∞ and τ on $|\bar{\rho}_0|$ is shown for the tensile deformation. First, figure 3(a) shows that e_∞ decreases as $|\bar{\rho}_0|$ increases. To interpret the result, we consider the definition of e_∞ from equation (5) which indicates that $e_\infty \propto 1/E_2$. Therefore, the increase in the density, which leads to decreased e_∞, implies an increase in E_2 or the overall stiffness of the system. Second, figure 3(b) shows that τ decreases with increased $|\bar{\rho}_0|$. From the definition of τ in equation (5), the term $(E_1 + E_2)/E_1E_2$ is not expected to change significantly and, therefore, $\tau \propto \eta$. This proportionality indicates that the increase in the density, which leads to decreased τ, results in lower η or lower damping capacity of the system. Combining the results from figures 3(a) and 3(b), we can see that the PFC model predicts high-density materials to exhibit high stiffness and low damping capacity. This prediction agrees in general with materials such as metals and ceramics [7].

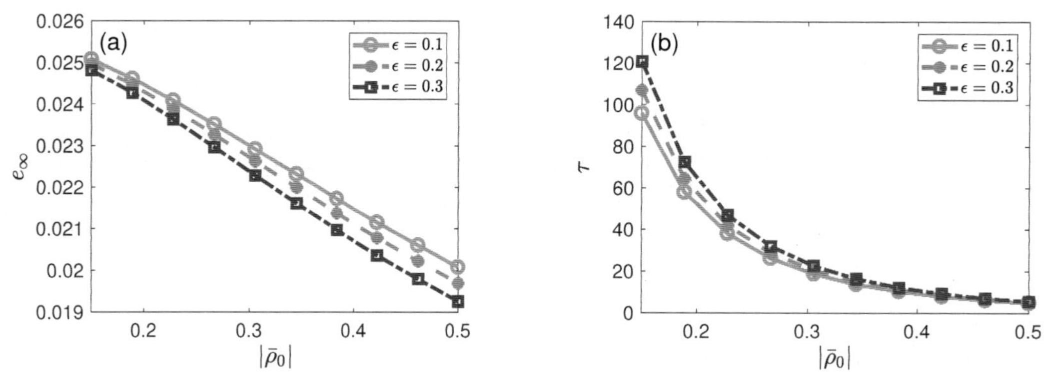

Figure 3. Plots of e_∞ and τ as the functions of $|\bar{\rho}_0|$ for the tensile deformation ($\Delta\bar{P} = 0.05\bar{P}_0$).

In figure 4, the dependence of e_∞ and τ on ϵ is shown for the tensile deformation. First, figure 4(a) shows that as ϵ decreases, e_∞ increases. We note that ϵ is related to the degree of undercooling from the solidification temperature [1]; therefore, a higher (lower) ϵ value corresponds to lower (higher) temperature. Then, the result from figure 4(a) indicates that increasing temperature (lowering ϵ) leads to higher e_∞ or lower stiffness of the system. Second, figure 4(b) shows that τ decreases with decreasing ϵ; however τ is less affected by ϵ for higher values of $|\bar{\rho}_0|$. This result indicates that, at lower density, increasing temperature (lowering ϵ) leads to lower damping capacity while at higher density, the damping capacity is less affected by temperature. Combining the results from figures 4(a) and 4(b), we find that the PFC model predicts, for lower-density materials, that higher temperature leads to lower stiffness and lower damping capacity. The decrease in materials stiffness with increased temperature is typically observed experimentally. Also, the decrease in damping capacity (faster strain response) at higher temperature agrees with what is generally observed in creep phenomena; in other words, the relaxation time in the creep phenomena generally increases with temperature [6]. Despite the agreement between the predictions from the PFC model and experimental observations, further developments need to be implemented in order to better understand the viscoelastic behavior from the PFC model; these improvements include extension to three-dimensional structures, more extensive parametric study, and investigation into other viscoelastic phenomena.

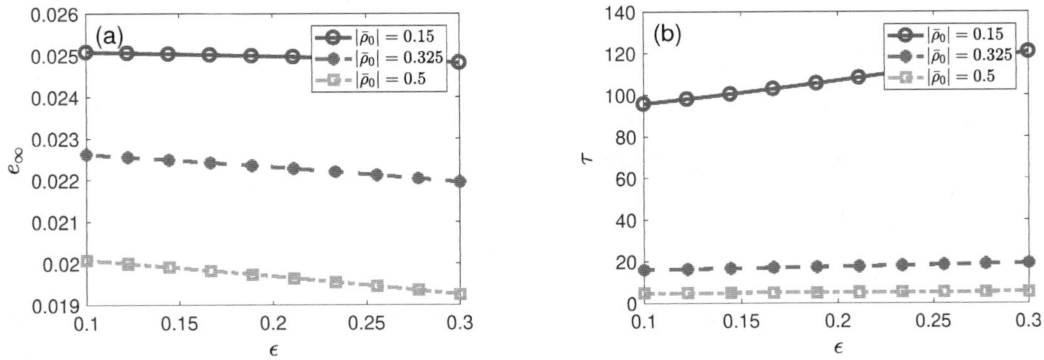

Figure 4. Plots of e_∞ and τ as the functions of ϵ for the tensile deformation ($\Delta\bar{P} = 0.05\bar{P}_0$).

5. Summary

In this work, we explored the viscoelastic creep behavior exhibited by the PFC model. We considered a one-dimensional crystalline solid subjected to step changes in pressure and the resulting time-dependent strain response was analyzed. From the parametric study, we found that the PFC model predicts higher-density materials to exhibit higher stiffness but lower damping capacity; this prediction agrees in general with materials such as metals and ceramics. Also, increasing temperature leads to lower material stiffness and damping capacity. This prediction agrees with experimental observations where materials typically become less stiff at high temperature and the relaxation time (proportional to the damping capacity) in creep phenomena generally decreases with temperature. The agreement between the PFC predictions and experimental observations shows promising capability of the PFC method to model viscoelastic phenomena. Nevertheless, further investigation is needed in order to better understand the viscoelstic behavior from the PFC model; future improvements include extension to three-dimensional structures, more extensive parametric study, and investigation into other viscoelastic phenomena.

Acknowledgments

This research was supported by the NSTDA-University-Industry Research Collaboration (NUI-RC) program under grant No. NUI-RC-M33-22-59-001D, Thailand Research Fund (TRF) under Grant No. TRG5880008, and High Performance Computing Services from (Thailand) National Electronics and Computer Technology Center (NECTEC). The authors are indebted to Dr.Sasawat Mahabunphachai for helpful discussion and support on the research.

References

[1] Elder K, Katakowski M, Haataja M and Grant M 2002 *Physical Review Letters* **88** 245701
[2] Emmerich H, Löwen H, Wittkowski R, Gruhn T, Tóth G I, Tegze G and Gránásy L 2012 *Advances in Physics* **61** 665–743
[3] Em-Udom J and Pisutha-Arnond N *(accepted) IOP Conference Series: Materials Science and Engineering*
[4] Wu K A and Karma A 2007 *Physical Review B* **76** 184107
[5] Elder K and Grant M 2004 *Physical Review E* **70** 051605
[6] Lakes R S 2009 *Viscoelastic Materials* (Cambridge University Press)
[7] Zhang J, Perez R and Lavernia E 1993 *Journal of Materials Science* **28** 2395–2404

IOP Conf. Series: Materials Science and Engineering **361** (2018) 012010 doi:10.1088/1757-899X/361/1/012010

Electrochemical redox reaction of zinc ions at a Ni electrode in $ZnCl_2$-based aqueous solutions

S Jeong

Department of Chemical Engineering, Soonchunhyang University, Asan, Chungnam 31538, Republic of Korea

E-mail: hamin611@sch.ac.kr

Abstract. The electrochemical redox reaction of zinc ions on a nickel electrode was investigated in aqueous solutions containing different concentrations of zinc chloride ($ZnCl_2$). A relatively good cycle efficiency was obtained from the high-concentration solution. The charge and discharge of the nickel electrode exhibited a high cycle efficiency of 92% in the fiftieth cycle in a 2 mol dm^{-3} (M) solution, and a low cycle efficiency of ~43% in the 0.02-M solution. Atomic force microscopy revealed that the dendritic zinc formed on Ni produced different morphologies as the electrolyte concentration increased, resulting in the changes in cycle efficiency. Moreover, the sizes of the zinc particles deposited on the nickel electrode were dependent on the electrolyte concentration. After five cycles, the zinc particles generated in the 0.02-M solution were smaller than those generated in the 2-M solution; however, after fifty cycles, the former particles were larger than the latter.

1. Introduction

Secondary batteries are devices that can be used semi-permanently by using an external power source and repeatedly cause electricity to flow through them in a direction opposite to discharge. Their importance is increasing with recent developments in the mobile communications industry, and they are expected to play a crucial role in the core industries of information, communication, and electronics, along with semiconductors and internet of things. This is because high-performance secondary batteries are expected to accelerate the development of the mobile communications industry. In order to realize high-performance and long-term use of electronic products, it is essential to lower the power consumption of the components and increase the capacity of the power sources. However, there is a limitation on the extent to which the power consumption of a device can be reduced; therefore, much effort has been focused on the development of a secondary battery with a high energy density and a long lifetime.

Since the characteristics of a battery are determined by the anode, cathode, and electrolyte, which are its main components, significant material development is required to improve the performance of a battery. One material that can meet this demand is zinc. It is a non-toxic metal that is abundant and has long been widely used as an anode material for primary cells. Zinc is inexpensive and has good electrochemical capacity (theoretically 820 Ah/kg). In recent years, especially as environmental issues have greatly increased, zinc is attracting more attention as an environmentally friendly material. For this reason, zinc electrodes are most widely used as negatively charged electrode materials for alkaline batteries, including Zn/air and Zn/Ni batteries [1-4].

Content from this work may be used under the terms of the Creative Commons Attribution 3.0 licence. Any further distribution of this work must maintain attribution to the author(s) and the title of the work, journal citation and DOI.
Published under licence by IOP Publishing Ltd

IOP Conf. Series: Materials Science and Engineering **361** (2018) 012010 doi:10.1088/1757-899X/361/1/012010

Although zinc exhibits excellent characteristics as an electrode material in a primary battery, in order to use it as an electrode in a secondary battery, the growth of dendritic zinc on the surface of the electrode caused by repeated charging and discharging must be overcome [5-8]. This dendritic zinc is one of the causes of irreversible loss of capacity of a battery, and when it grows, the lifetime is shortened and safety deteriorates. Therefore, many efforts have been made to suppress the formation of dendritic zinc, but no effective method has yet been found.

There have been extensive studies on dendritic lithium metal produced by the dissolution and deposition of metal ions [9-12]. This is because a lithium secondary battery using the redox reaction of lithium ions as an electrode reaction exhibits excellent performance. Recently, interesting research results have been reported for the dissolution and deposition of lithium. The redox reaction of lithium ions occurring at the interface between graphite (or nickel metal) electrodes and propylene carbonate-based solutions containing lithium salt is greatly influenced by the electrolyte concentration. Even without the use of special additives, changes in the electrolyte concentration significantly change the reaction in the battery. Lithium insertion into graphite does not occur in a low-concentration solution but does occur in a high-concentration solution, and the growth of dendritic lithium on the nickel surface is greatly suppressed in a high-concentration solution.

In this study, the findings obtained from the redox reaction of lithium ions were applied to an electrochemical system using zinc as an electrode. The aim of this study was to understand the effect of zinc ion concentration on the electrochemical redox reaction of zinc by focusing on the concentration of zinc chloride dissolved in the electrolyte solution as a factor affecting the growth of dendritic zinc.

2. Experimental

The charge/discharge measurements and cyclic voltammetry (CV) were performed using a laboratory-made cell, as shown in figure 1. A nickel plate (20 × 20 × 0.5 mm, Nilaco Co.) polished with alumina powder (0.1 μm, Bailkowski International Co.) to a mirror finish was used as the working electrode. Cleanly polished nickel was brought into contact with an electrolyte solution using an O-ring. A platinum mesh and Ag/AgCl (99.99%) were used as the counter and reference electrodes, respectively. The electrolyte solution was prepared by dissolving $ZnCl_2$ (95%, Junsei Chemical Corp.) in deionized water (HPLC, Burdick and Jackson). The current density for the zinc dissolution and deposition was set to 0.4 mA cm^{-2}. Zinc was deposited at 0.6 C cm^{-2} and dissolved until the potential reached −0.19 V versus Ag/AgCl. CV was conducted between −1.0 and 0.0 V at a sweep rate of 0.5 mV s^{-1}. Atomic force microscopy (AFM) observations were carried out in conventional contact mode using an AFM system (PicoSPM, Molecular Imaging). Pyramidal silicon nitride (Si_3N_4) tips mounted on a triangular micro-fabricated cantilever with a length of 200 μm and a spring constant of 0.02 N m^{-1} (OMCL-TR400PSA, Olympus) were used for the AFM measurements.

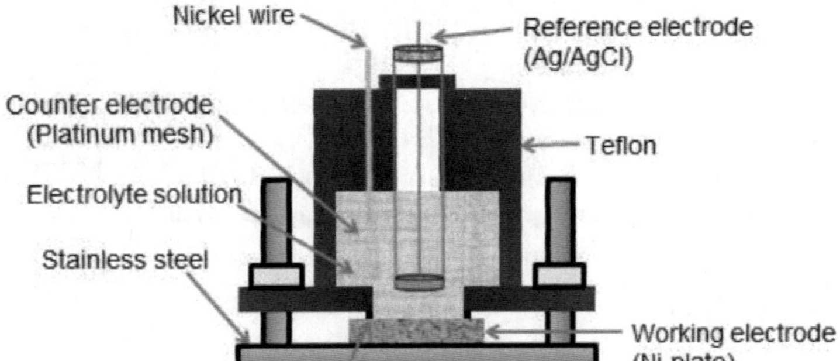

Figure 1. Three-electrode cell used in electrochemical experiments.

IOP Conf. Series: Materials Science and Engineering **361** (2018) 012010 doi:10.1088/1757-899X/361/1/012010

3. Results and discussion

The cycle efficiencies for the electrochemical redox reaction of zinc in the three electrolytic solutions of different concentrations are shown in figure 2. The cycle efficiency is defined as the ratio of electricity used for charging (zinc deposition) to that used for discharging (zinc dissolution) in each cycle. In a 0.02 mol dm^{-3} (M) solution, the variation in efficiency according to the cycle progression was very large and the cycle efficiency decreased sharply after ten cycles, resulting in ~43% cycle efficiency in the fiftieth cycle. In contrast, in the 0.2- and 2-M solutions, the cycle efficiency increased sharply in the initial five cycles, and then was stable up to the fiftieth cycle. In all three electrolyte solutions used, the efficiencies began at a very low value of 40–80% in the first cycle and then increased in the first five cycles. This is probably due to the irreversible consumption of charge during stabilization of the electrode by the film generated on the Ni electrode from the decomposition of the electrolyte solution.

The different electrochemical behaviors of the Ni electrode for the different solutions as described above is presumed to be because the solvation structure of the zinc ions present in the electrolyte is different in each solution. In the electrolyte solution, the zinc ions do not exist alone, but in a solvated state that is weakly bound to the solvent, and the solvation number is known to depend on the concentration of the electrolyte. Therefore, the structure of the solvated zinc ions affects the deposition and dissolution of zinc, which occurs as a result of the electron transfer reaction between the solvated zinc ions and the electrode, which may have caused the difference in efficiency.

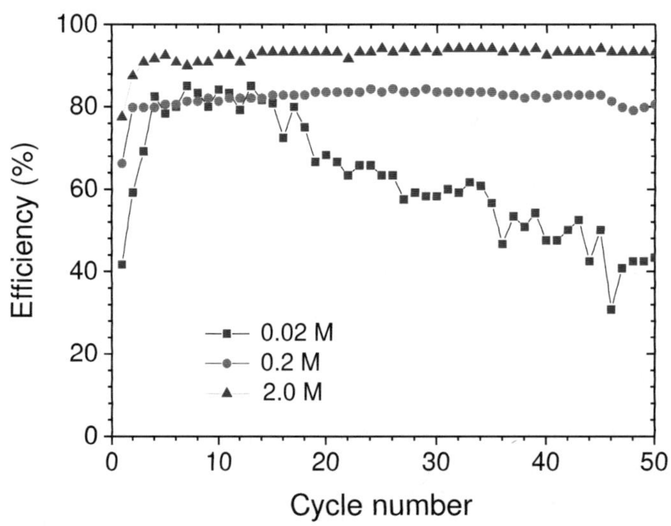

Figure 2. Cycling efficiencies of the Ni electrode in various $ZnCl_2$-based electrolytes with different concentrations.

The results shown in figure 2 indicate that the electrochemical reaction in the nickel electrode is changing with each cycle. In order to confirm this, the CVs of the electrodes after fifty cycles in the 0.02- and 2-M solutions were measured and the results are shown in figure 3. The redox peaks near −1.0 V due to deposition and dissolution of zinc were observed in the 2-M solution but not in the 0.02-M solution. That is, there is no doubt that the state of the surface of the Ni electrode was changed by the progress of the cycle, and thereby the deposition and dissolution reactions of zinc were affected.

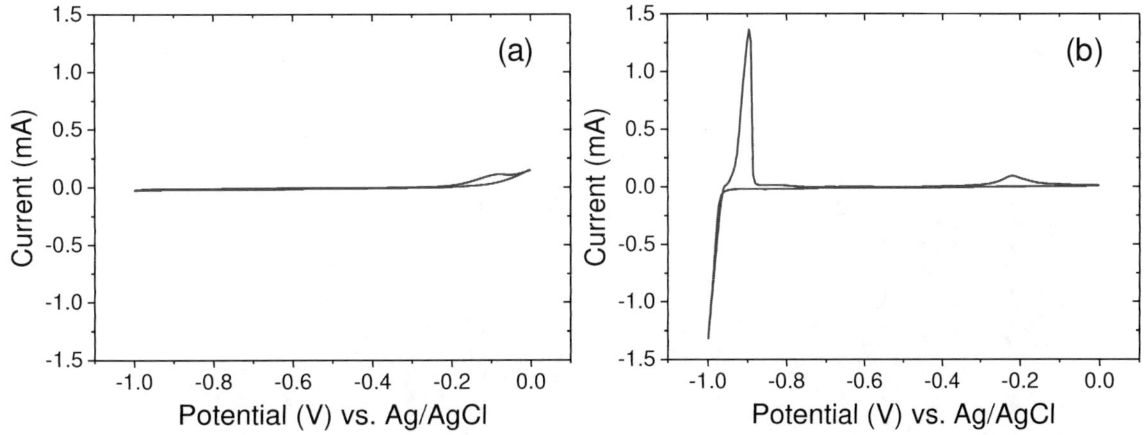

Figure 3. Cyclic voltammograms of Ni electrodes in the (a) 0.02-M and (b) 2-M solutions after the fiftieth cycle.

The above cycle efficiency and CV results indicate that the growth of dendritic zinc is highly dependent on the electrolyte concentration. In order to confirm the relationship between the growth of dendritic zinc on the electrode surface and the electrolyte concentration, the surface of the electrode cycled in the 0.02-M solution with poor cycle efficiency and in the 2-M solution with relatively good cycle efficiency was observed using AFM. In order to observe the surface of a sample using AFM, a substrate with a very flat surface is required; the surface of the Ni electrode should be very flat and is therefore sufficient for AFM observations. Figure 4 shows the surface morphology of Ni metal polished using alumina as a substrate used for zinc deposition and dissolution. A linear scratch formed by the alumina during polishing was observed over the entire surface, and it was found that it had sufficient roughness to be observed with AFM.

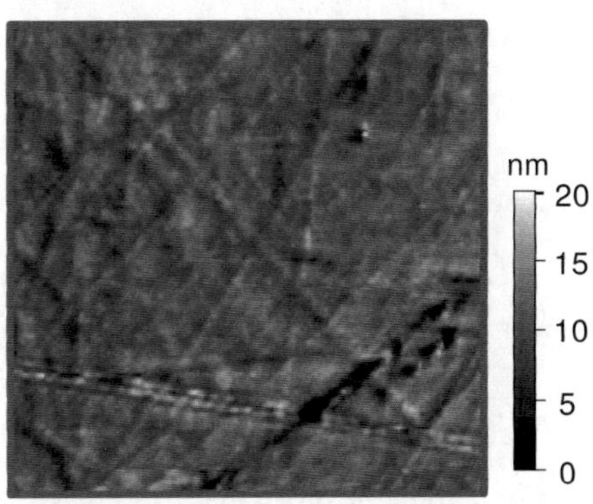

Figure 4. AFM image (5 × 5 μm) of the pristine Ni substrate polished with alumina powder

Figure 5 shows AFM images obtained from the surface of the electrode after five, ten, thirty, and fifty cycles in the 0.02-M electrolyte solution. Scratch marks due to polishing are still observed on the surface of the electrode after five cycles (figure 5(a)), and substances presumed to be dendritic zinc are observed all over the surface. These substances grew to a fairly large size after 30 cycles, with very large variations in size (figure 5(c)). A clear image was not obtained after ten cycles (figure 5(b)), which is believed to be due to the very irregular growth of the dendritic zinc on the surface, resulting in poor surface smoothness. That is, at ten cycles, the growth of dendritic zinc has already begun. This is consistent with the rapid decrease in cycle efficiency after ten cycles, as shown in figure 2. By thirty cycles, dendritic lithium had grown over the entire surface, and after fifty cycles the flatness of the electrode surface again became very poor, and no clear surface morphology was obtained.

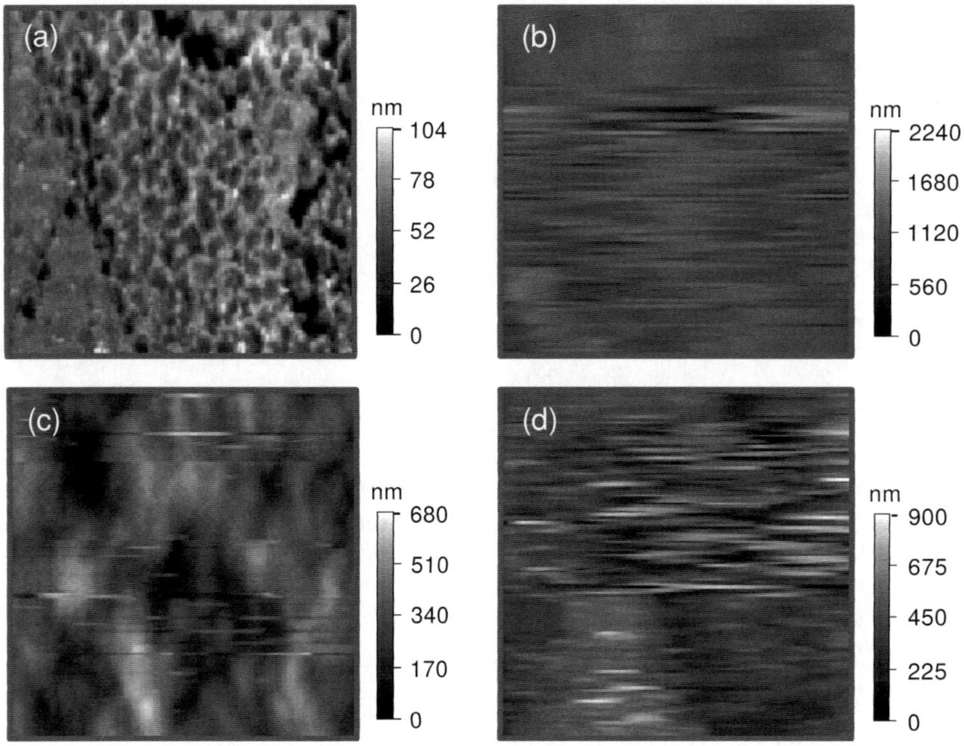

Figure 5. AFM images (5 × 5 μm) of the Ni electrode in a 0.02-M electrolyte solution after (a) five, (b) ten, (c) thirty, and (d) fifty cycles.

The above AFM results revealed that the growth of dendritic zinc was very rapid in the 0.02-M solution, and progressed very irregularly according to the location of the electrode. Figure 6 shows AFM images of the electrodes in the 2-M solution under the same conditions as figure 5 for comparison. Scratches similar to those observed in the 0.02-M solution were observed on the surface of the electrode after five cycles, as shown in figure 6(a), and particle-like substances were generated on the surface. The larger number of scratches in figure 6(a) than in figure 5(a) after five cycles indicates that a smaller amount of dendritic zinc was formed on the electrode surface in the 2-M solution. That is, it can be considered that the growth of dendritic zinc was suppressed in the 2-M solution. However, after repeated cycles in the 2-M solution, the amount of dendritic zinc increased. After fifty cycles, scratches due to polishing are not observed at all), and the sizes of the particle-like substances deposited on the surface are slightly increased (figures 6(b)–(d). In addition, the surface of the electrode obtained from the 2-M solution was very flat, and thus relatively clear AFM images could be obtained. From the AFM results shown in figures 5 and 6, it is necessary to uniformly deposit small-sized zinc particles on the electrode surface in order to suppress the growth of dendritic zinc, which is closely related to the electrolyte concentration.

4. Conclusions

The electrochemical redox reaction of zinc is strongly influenced by electrolyte concentration. The dendritic zinc produced in 0.02- and 2-M solutions exhibited different shapes in the initial cycles. As the number of cycles increased, the dendritic zinc in the 2-M solution was maintained without significant changes in shape, while dendritic lithium in a 0.02-M solution has been observed to grow significantly. The effect of electrolyte concentration is presumed to be due to the structural differences in solvated zinc ions. However, from this study, it is unclear what kind of difference this is, and further analysis by Raman spectroscopy is necessary.

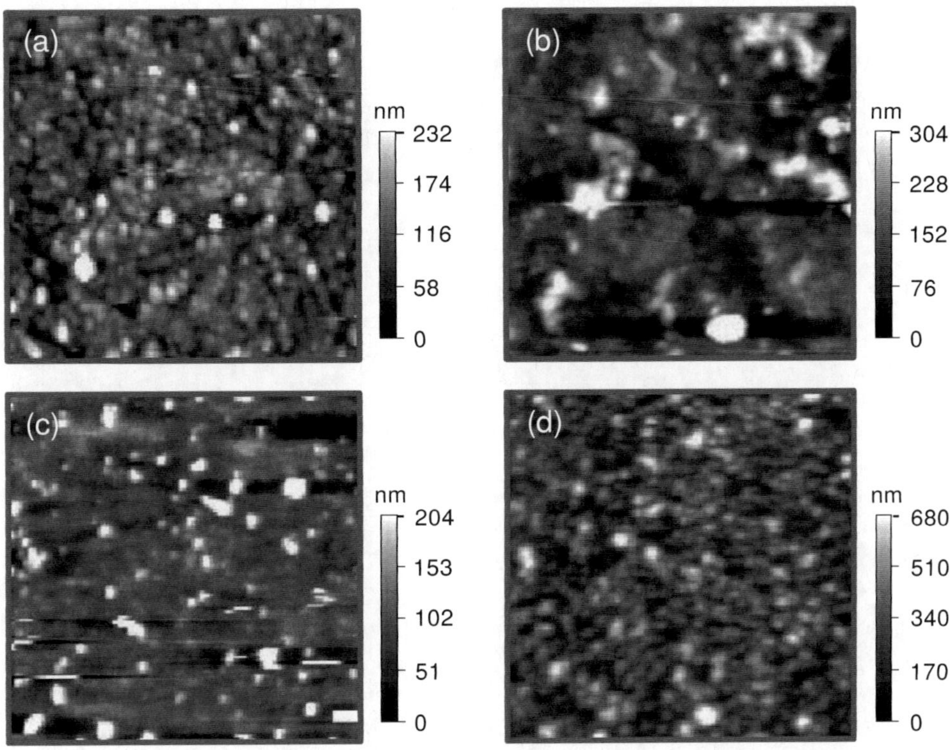

Figure 6. AFM images (5 × 5 μm) of the Ni electrode in the 2-M electrolyte solution after (a) five, (b) ten, (c) thirty, and (d) fifty cycles.

References

[1] Shivkumar R, Kalaignan G P and Vasudevan T 1998 *J. Power Sources* 5 90
[2] Chang H and Lim C 1997 *J. Power Sources* 66 115
[3] Jindra J 2000 *J. Power Sources* 88 202
[4] Kannan A, Muralidharan S, Sarangapani K, Balaramachabdran V and Kapali V 1995 *J. Power Sources* 57 93
[5] Kim H, Popov B and Chen K *2003 Corros. Sci.* 45 1505
[6] Veeraraghavan B, Haran B, Prabhu S and Popov B 2003 *J. Electrochem. Soc.* 150 B131
[7] Lee C, Sathiyanarayanan K, Eom S and Yun M 2006 *J. Power Sources* 160 1436
[8] Lan C, Lee C and Chin T 2007 *Electrochim. Acta* 52 5407
[9] Wrodnigg G H, Besenhard J O and Winter M 1999 *J. Electrochem. Soc.* 146 470
[10] Jeong S K, Inaba M, Mogi R, Iriyama Y, Abe T and Ogumi Z *Langmuir* 17 8281
[11] Jeong S K, Inaba M, Iriyama Y, Abe T and Ogumi Z *Electrochim. Acta* 148, A989
[12] Jeong S K, Inaba M, Iriyama Y, Abe T and Ogumi Z *Electrochem. Solid State Lett.* 6 A13

Acknowledgments
This work (Grants No.C0502212) was supported by Business for Cooperative R&D between Industry, Academy, and Research Institute funded Korea Small and Medium Business Administration in 2017.

IOP Conf. Series: Materials Science and Engineering **361** (2018) 012011 doi:10.1088/1757-899X/361/1/012011

A green membrane distillation system for seawater desalination: Response surface modelling and optimization

Siti Khadijah Hubadillah[a], Mohd Hafiz Dzarfan Othman[a*], Paran Gani[b], Ahmad Fauzi Ismail[a], Mukhlis A. Rahman[a], Juhana Jaafar[a]

[a]Advanced Membrane Technology Research Centre (AMTEC), Universiti Teknologi Malaysia, 81310 Skudai, Johor, Malaysia.
[b]Faculty of Civil Engineering and Earth, Universiti Tun Hussein Onn Malaysia, 84600 Parit Raja, Batu Pahat, Johor.

Email: hafiz@petroleum.utm.my

Abstract. In this work, response surface methodology (RSM) was applied for modelling and optimization of operating parameters for water desalination by direct contact membrane distillation (DCMD) system using cost effective and green ceramic hollow fibre membrane prepared from rice husk waste. Operating parameters including feed temperature, permeate temperature and feed flow rate were selected and the optimum parameters were determined for DCMD salt rejection. The developed model for permeate flux response was statistically validated by analysis of variance (ANOVA) which showed a high value coefficient of determination value (R^2 = 0.9954). In this study, the obtained optimum operating parameters were found to be 70.15℃ of feed temperature, 19.84℃ permeate temperature and 0.58 LPM of feed flow rate.

1. Introduction

"Fresh water is the world's first and foremost medicine". The world is poised at the brink of a severe global crisis especially lack of fresh water. Accordingly, seawater as alternative source for fresh water has been explored due to its large percentage (>97%) of total water available on earth, leading increasing number of studies on seawater desalination. Among all technologies, membrane distillation (MD) system has received increasing attention, especially after the rapid advances in MD membranes and solar energy. There are four main configuration in MD [1]; direct contact membrane distillation (DCMD), air gap membrane distillation (AGMD), sweeping gas membrane distillation (SGMD), and vacuum membrane distillation (VMD). Consequently, DCMD is the most studied configuration because the condensation step is carried out inside the membrane module leading to a simple operation mode without the need of external condensers like in SGMD and VMD configurations. In DCMD configuration, the hot feed solution is in direct contact with the inlet membrane side surface. Therefore, evaporation takes place at the feed membrane surface. Vapor is moved by the pressure difference across the membrane to the permeate side and condenses inside membrane module [2]. Because of the hydrophobic nature of the membrane, the liquid feed cannot penetrate inside dried membrane pores. Up to the present, DCMD has been applied, in major part, for the production of fresh water, in wastewater treatment and reuse and in the food industry.

Content from this work may be used under the terms of the Creative Commons Attribution 3.0 licence. Any further distribution of this work must maintain attribution to the author(s) and the title of the work, journal citation and DOI.
Published under licence by IOP Publishing Ltd

The major part of the reported researches on membrane distillation was carried out using conventional method of experimentation, changing one of the independent parameters while maintaining the others at fixed levels. Conventional method involves an elevated number of experimental runs, which are time consuming, expensive and ignore the interaction effects between the dominant parameters affecting the performance of the membrane distillation process. Research Surface Methodology (RSM) has been applied successfully in scientific and research areas such as chemistry, physics, biochemistry, biology, environment, etc. In the field of membrane distillation, few studies have been carried out using RSM.

Khayet et al. [3] used the RSM for the optimization of DCMD process together with prediction of permeate fluxes when using NaCl salt aqueous solution. The considered factors were the stirring velocity, the feed temperature and the NaCl concentration. It was found that the DCMD parameters affect the permeate flux in the same way as observed from the classical method of experimental flux. There is a good agreement between the experimental and the predicted permeate fluxes by RSM. However, the use of polymeric membranes, like PVDF, is severely restricted for the long term operation at elevated temperatures or drastic chemical conditions. Thus, ceramic membranes offer several advantages over polymeric membranes especially due to its non-swelling behaviour and thermal stability, which is desirable in MD application [4-6]. In our previous work [7], we have successfully prepared a cost effective and green ceramic hollow fibre membrane derived from rice husk waste and prepared through phase inversion and sintering technique. In was revealed that a very highest flux more than 35 kg/m^2h (data in review), comparable to the polymeric one, could be obtained. Unfortunately, there is some drawbacks using ceramic membrane in MD which is its poor wettability properties. Other than its hydrophobicity/hydrophilicity effect, other major reason is due to the microfiltration pores of ceramic membrane that can lead salt in feed solution to pass through. Thus, the objectives of this study is to model and optimize the DCMD parameters using green ceramic hollow fibre membrane prepared from rice husk waste towards salt rejection performance.

2. Materials and Methods

2.1. DCMD experimental setup

Experiments were carried out using a green ceramic hollow fibre membrane (CHFM) made of rice husk waste that converted into silica through calcination process. Then, ceramic suspension containing silica based rice husk ash as main material, polyethersulfone (PESf) as binder and 1-methyl-pyrrolidone (NMP) as solvent. Tap water was used as both external and internal coagulation bath. After that, sintering process was conducted to produce final CHFM. The detail preparation of the CHFM was reported in our previous work [7]. Accordingly, the prepared CHFM was modified from its natural hydrophilicity into hydrophobic through FAS grafting process [8].

2.2. Experimental design and statistical analysis

In this study, the Central Composite Design (CCD) was applied for the optimization of DCMD process used for desalination of NaCl salt solution. This is due to the design that is suitable for fitting a quadratic surface and helps to optimize the effective parameters with minimum number of experiments, at the same time, analyzing the interaction between parameters. Three main parameters in DCMD have been chosen which are feed temperature, permeate temperature and feed flow rate, to study their interaction towards better salt rejection. In CCD, DCMD parameters and performances were termed as variables and responses, respectively. Consequently, mathematical function is assumed for the response in terms of the significant independent variables. A quadratic model corresponding to the following second order equation was built to describe the response [9]:

$$Y = b_0 + \sum_i b_i X_i + \sum_i b_{ii} X_i^2 + \sum_{ij} b_{ij} X_i X_j \qquad (1)$$

where Y is the response, b_o the constant coefficient, b_i the linear coefficients, b_{ii} the quadratic coefficients, b_{ij} the interaction coefficients and X_i, X_j the coded values of the variables.

The statistical significance of the models was justified through analysis of variance (ANOVA) for polynomial model with 95% confidence level, and residual plots were used to examine the goodness of models fit. The quality of the fit polynomial model was also expressed by the coefficient of determination R^2. In this study, three factors including feed temperature, permeate temperature and feed flow rate with five levels were employed for response surface modeling and optimization of DCMD process. According to CCD, a total number of 32 experiments have been performed with 3 replicate points to estimate the experimental error.

3. Results and Discussions

Among all desalination technologies, direct contact membrane distillation (DCMD) is a recent method that received worldwide attention due to its excellent desalination performance at cost which appears to be reasonable compared to reverse osmosis (RO) membrane [10]. In addition, newly-developed green ceramic membrane from waste rice husk was used in DCMD, for the first time, for desalination application. In DCMD process, the hot solution (feed) is in direct contact with the hot membrane side surface. Therefore, evaporation takes place at the feed-membrane surface. The vapour is moved by the pressure difference across the membrane to the permeate side and condenses inside the membrane module. Accordingly, an optimum parameters of the process such as feed temperature, permeate temperature and feed flow rate should be investigated to produce high water vapor flux. An optimization using Research Surface Methodology (RSM) method has been performed. Table 1 presents the regression model developed for the prediction of salt rejection using analysis of variance (ANOVA). It was observed that the model F-value of 20.67 implies the model is significant. In addition, value of Prob > F less than 0.0500 indicate that the model are significant. Therefore, in this case, C and C^2 which are both represent feed flow rate are significant model terms. The "lack of fit F-value" is 1.23 implies the lack of fit is not significant to the pure error There is a 33.85% chance that a "lack of fit F-value" this large could due to noise. Based on Myers et al., non-significant lack of fit presents a good model, which is obtained for this work [11]. It should be mentioned that the p-value obtained for this work is less than 0.05 (p-value = <0.0001). This means that the obtained salt rejection model is highly significant and the regression model gives a good prediction of the experimental data. The coefficient of determination (R^2), which is the proportion of variation in the response attributed for the regression model, should be close to 1. The R^2 of the regression model is 0.9954, meaning that more than 99.54% of the data deviation can be explained by the empirical model, which indicates that the regression model is statistically significant. Moreover, it is preferred to use the adjusted R^2_{Adj} values to evaluate the model adequacy since it is adjusted for the number of terms in the model. The R^2_{Adj} value is 95.1%, indicating a high degree of correlation between the experimental and predicted responses.

Table 1. Anova for the response surface quadratic model

Sources	Sum of squares	df	Mean square	F- Value	P-Value
Model	106.90	9	11.87732399	20.67	< 0.0001
A-Feed Temperature	0.00	1	0	0.00	1.0000
B-Permeate Temperature	0.00	1	0.002857143	0.00	0.9444
C-Feed Flow Rate	66.88	1	66.88285714	116.38	< 0.0001
AB	0.72	1	0.72	1.25	0.2751
AC	0.00	1	0	0.00	1.0000
BC	0.00	1	0.005	0.01	0.9265
A^2	0.67	1	0.671836228	1.17	0.2913

B^2	0.67	1	0.671836228	1.17	0.2913
C^2	29.77	1	29.76516956	51.79	< 0.0001
Residual	12.64	22	0.574674277		
Lack of Fit	3.36	5	0.671233487	1.23	0.3385
Pure Error	9.29	17	0.54627451		
Cor Total	119.54	31			
R^2= 0.9954, Adj. R^2= 0.9510, Pred. R^2= 0.69.23, Adequate Precision=12.562					

Figure 1 shows the 3D response curves that presenting the interaction of the variables and optimum level of each variables and optimum level of each variable for maximum response. Figure 1(A) shows the response of salt rejection with varying permeate temperature and feed temperature. As can be seen, both factors have a slight effect on DCMD salt rejection. At center value of PT and FT a maximum response value was obtained, the salt rejection is 100% as calculated by regression model. Whereas, Figure 1(B) shows the responses of DCMD salt rejection with varying FT and Q_F. It is observed that the effect of feed flow rate (Q_F) on salt rejection is highly significant than the effect of feed temperature (FT). This is highly due to the pore size of ceramic membrane that in microfiltration range [12]. The combined effects of PT and Q_F on salt rejection is shown in Figure 1(C). As expected, the PT do not give any effect on salt rejection. As a result, similar trend also obtained by the effect of feed flow rate on salt rejection. Therefore, it was revealed that the feed flow rate (Q_F) has the major effect towards DCMD salt rejection.

To confirm the model adequacy and the validity of optimization procedure, experiments were carried out under optimized conditions and the results were compared with the predicted values of response using the model equation. The experiments were conducted in triplicates and mean values are reported. The optimization reveals that to obtain the highest 100% of salt rejection using hydrophobic ceramic membrane made from rice husk waste, the three main parameters in DCMD which are feed temperature, permeate temperature and feed flow rate should set at 70.15℃, 19.84℃ and 0.58 LPM, respectively. These results represent 0.5% as deviation between experimental (100%) and predicted values (100.519%). In fact, the highest salt rejection cannot be more than 100%, which means that the experimental value is the best value obtained. This indicates the suitability of the developed regression model and it may be noted that these optimal values are valid within the specified range of DCMD process variables.

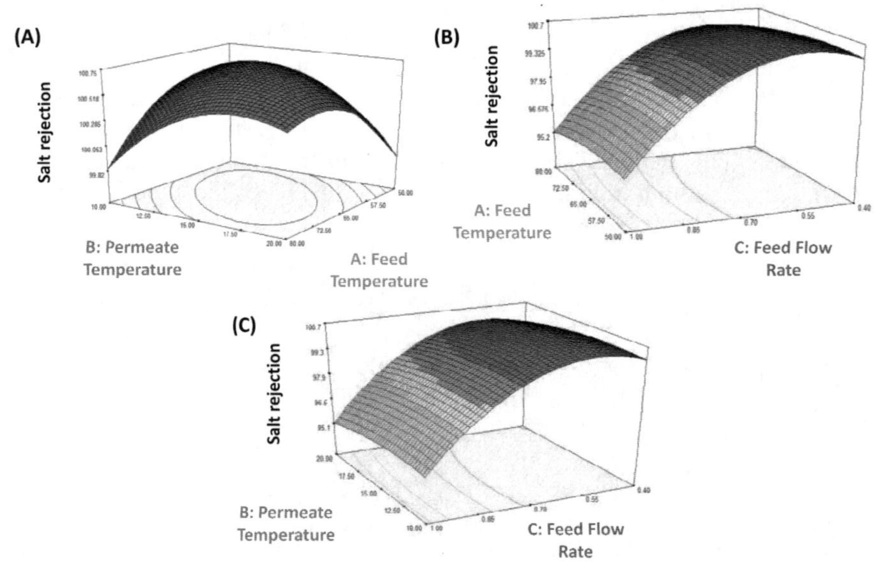

Figure 1. 3D response surface plot of salt rejection as function of operating factors; (A) feed temperature vs permeate temperature, (B) feed temperature vs feed flow rate, and (C) permeate temperature vs feed flow rate

4. Conclusions

This study investigates the modeling and optimization of lab scale DCMD for aqueous NaCl salt solution for the purpose of seawater desalination application using experimental design and response surface methodology. Effects of operating conditions which are feed temperature, permeate temperature and feed flow rate, were studied to optimize DCMD salt rejection using ceramic membrane made from rice husk waste. As a result, an optimization has been obtained by setting the feed temperature, permeate temperature and feed flow rate of 70.15°C, 19.84°C and 0.58 LPM, respectively. The response surface model for DCMD permeate flux was statistically validated by ANOVA and provided good quality to predict the response, whose $R^2 = 0.9954$ and $R^2_{Adj} = 0.9510$. It was also found that the feed flow rate giving the most significant effect towards DCMD salt rejection.

References

[1] González D, Amigo J and Suárez F 2017 *Renewable and Sustainable Energy Reviews* **80** 238.
[2] Cheng D, Gong W and Li N 2016 *Desalination* **394** 108.
[3] Khayet M, Cojocaru C and García-Payo C 2007 *Industrial & Engineering Chemistry Research* **46** 5673.
[4] Hubadillah SK, Othman MHD, Harun Z, Ismail AF, Iwamoto Y, Honda S, Rahman M A, Jaafar J, Gani P and Mohd Sokri M N 2016 *Ceramics International* **42** 15547.
[5] Hubadillah SK, Harun Z, Othman MHD, Ismail AF and Gani P 2016 *Journal of Asian Ceramic Societies* **4** 164.
[6] Hubadillah SK, Othman MHD, Harun Z, Ismail AF, Rahman MA, Jaafar J, Jamil SM and Mohtor NH 2017 *Materials Letters* **191** 119.
[7] Hubadillah SK, Othman MHD, Harun Z, Ismail AF, Rahman MA and Jaafar J 2017 *Ceramics International* **43** 4716.
[8] Picard C, Larbot A, Tronel-Peyroz E and Berjoan R 2004 *Solid State Sciences* **6** 605.
[9] Boubakri A, Hafiane A and Bouguecha SAT 2014 *Journal of Industrial and Engineering Chemistry* **20** 3163.
[10] Pangarkar BL, Sane MG and Guddad M 2011 *ISRN Materials Science* **9** 1.
[11] Myers RH, Montgomery DC and Anderson-Cook CM, *Response Surface Methodology: Process and Product Optimization Using Designed Experiments* Wiley United States of America.
[12] Wang S, Kang Z, Xu B, Fan L, Li G, Wen L, Xin X, Xiao Z, Pang J, Du X and Sun D 2017 *Inorganic Chemistry Communications* **82** 64.

Acknowledgement

The authors gratefully acknowledge the financial support from the Ministry of Education Malaysia under the Higher Institution Centre of Excellence Scheme (Project Number: R.J090301.7846.4J192), Universiti Teknologi Malaysia under the Research University Grant Tier 1 (Project number: Q.J130000.2546.12H25), Nippon Sheet Glass Foundation for Materials Science and Engineering under Overseas Research Grant Scheme (Project number: R.J130000.7346.4B218) and Kurita Water and Environment Foundation under KWEF Research Grant Program.

Vibration suppression with electromagnetic hybrid vibration isolators

Fan Chen

School of Mechanical Science and Engineering, Huazhong University of Science and Technology, Wuhan, China

Email: chenf04@163.com

Abstract. The design and control of a novel electromagnetic hybrid vibration isolator is presented in this paper. In the proposed hybrid vibration isolator configuration, an electromagnetic actuator and a rubber vibration isolator are configured in a parallel structure with the rubber vibration isolator installed inside the mover of the actuator. The magnetic force of the actuator is linearized by magnetic configuration design. Filtered-x least mean square (FxLMS) control algorithm is implemented for hybrid vibration isolation. Experimental results show that the vibration is suppressed significantly with the proposed hybrid vibration isolator and the FxLMS control algorithm.

1. Introduction

Large rotary machines, like diesel engines and large motors, always cause severe vibration problems and induce a lot of noise. Passive rubber vibration isolators[1], are normally used for vibration isolation of these kinds of large rotary equipments. However, for low frequency vibration isolation, the performance of passive isolators are quite limited. Hybrid vibration isolators combine passive isolators and actuators together, and use active control laws to suppress the vibrations. Comparing to passive isolators, the hybrid vibration isolator can not only damp the high frequency vibrations, but also obtain better performance in the low frequency region. In a hybrid vibration isolator, the passive isolator and the actuator can be structured either in series or in parallel.

For hybrid vibration isolators structured in series, Chen and Shih[2] presented a hybrid pneumatic vibration isolator, in which an active pneumatic actuator and a passive pneumatic vibration isolator were connected in series. The passive pneumatic isolator was used to support the gravity of the payload, while the active pneumatic actuator was used to suppress the low frequency disturbances. However, the size of the proposed hybrid vibration isolator is too large to be implemented. Gao et al.[3] proposed a compact hybrid vibration isolator structure, in which a piezoelectric actuator and a passive rubber vibration isolator were integrated in series. They reported a 10dB reduction in the vibration transmissibility over the purely passive ways. But for piezoelectric actuators, the stroke is only several dozens of micrometers, thus restricting the vibration isolation performance[4].

For hybrid vibration isolators configured in parallel, Liu[5] built a hybrid vibration isolation system in which a voice coil actuator worked in parallel with four springs. Ahn et al.[6] installed an electromagnetic actuator in parallel with three pneumatic vibration isolators, and built a hybrid vibration isolation system. The work in [5] and [6] demonstrated that the parallel structured hybrid vibration isolation system can achieve a good vibration isolation performance. But the proposed

Content from this work may be used under the terms of the Creative Commons Attribution 3.0 licence. Any further distribution of this work must maintain attribution to the author(s) and the title of the work, journal citation and DOI.
Published under licence by IOP Publishing Ltd

IOP Conf. Series: Materials Science and Engineering **361** (2018) 012012 doi:10.1088/1757-899X/361/1/012012

configurations take up too much space. To be more practical for implementation, the actuators and the passive vibration isolators need to be integrated more compactly as one vibration isolator. Daley et al.[7] presented a compactly designed hybrid vibration isolator in which an electromagnetic actuator and several springs are structured in parallel. Li et al.[8] proposed to integrate an electromagnetic actuator inside a pneumatic vibration isolator to develop a new type of compact hybrid vibration isolator. However, the electromagnetic actuators in [7, 8] have nonlinear force-current relations, which complex the controller design. Also, integrating the electromagnetic actuator inside the passive vibration isolator restricts the actuator size as well as the maximum magnetic force output, therefore, limits the achievable vibration isolation performance.

This paper presents a novel parallel configured electromagnetic hybrid vibration isolator, in which the electromagnetic actuator is designed to have a linear relationship between the magnetic force and the coil current, which can simplify the controller design process. In the proposed hybrid vibration isolator structure, a passive rubber vibration isolator is installed inside the electromagnetic actuator. Therefore, the actuator size is not limited by the passive isolator, and can be designed to generate larger magnetic force which allows the hybrid vibration isolator to achieve a better performance. FxLMS control algorithm is implemented for hybrid vibration isolation. Experimental results show that significant vibration reduction is achieved with the proposed hybrid vibration isolators and the FxLMS control algorithm.

2. Electromagnetic hybrid vibration isolator design

2.1. Configuration of the hybrid vibration isolator

The configuration of the proposed hybrid vibration isolator is shown in Figure 1 (a) and the fabricated prototype hybrid vibration isolator is shown in Figure 1 (b). The designed hybrid vibration isolator is mainly composed of a rubber vibration isolator, an electromagnetic actuator, a base, a safety plate and a connector. The electromagnetic actuator includes a moving part and a stationary part. The mover of the actuator comprises armature cores and permanent magnets. The stator of the actuator includes stator cores and a coil which are boned together with epoxy. The electromagnetic actuator and the rubber vibration isolator are connected in parallel, and the rubber vibration isolator is installed inside the mover of the actuator. The stator of the actuator is fixed to the base. The connecter is used for installing the hybrid vibration isolator underneath the vibrating structure. The aluminum safety plate and the rubber pad are used to protect the mover of the actuator from collision during the vibration isolation process. An accelerometer is glued to the base of the hybrid vibration isolator to measure the acceleration signals of the base for real time control.

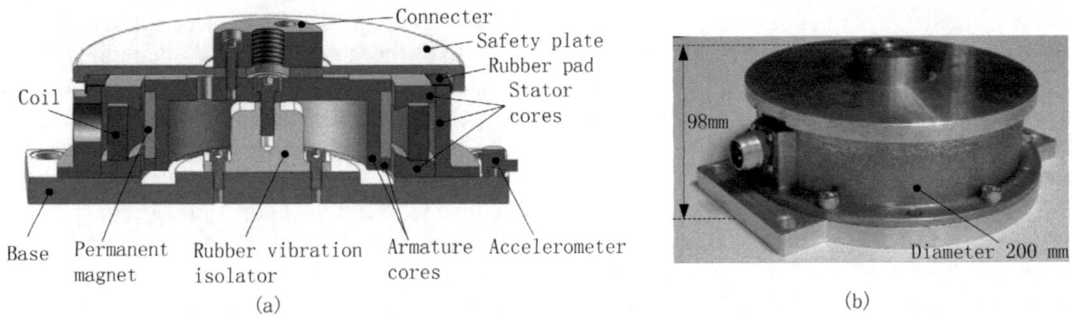

(a) (b)

Figure 1. Configuration of the hybrid vibration isolator, (a) 3D model; (b) Fabricated prototype.

2.2. Working principle of the electromagnetic actuator

The magnetic configuration of the proposed electromagnetic actuator is shown in Figure 2. The coil and the stator cores work as an electromagnet when the coil is supplied with current. If the current direction in the coil and the magnetization direction of the permanent magnet are the same as the ones

IOP Conf. Series: Materials Science and Engineering **361** (2018) 012012 doi:10.1088/1757-899X/361/1/012012

shown in Figure 2, the directions of the magnetic flux flowing through the stator cores will be clockwise. Then, a north pole and a south pole will be generated at the lower and upper sides of the stator, respectively. The mover of the actuator will sense a repelling force at the lower air gap and an attractive force at the upper air gap. Thus, a net force in the positive y direction will be generated on the mover. If the current direction reverses, the force direction will change accordingly.

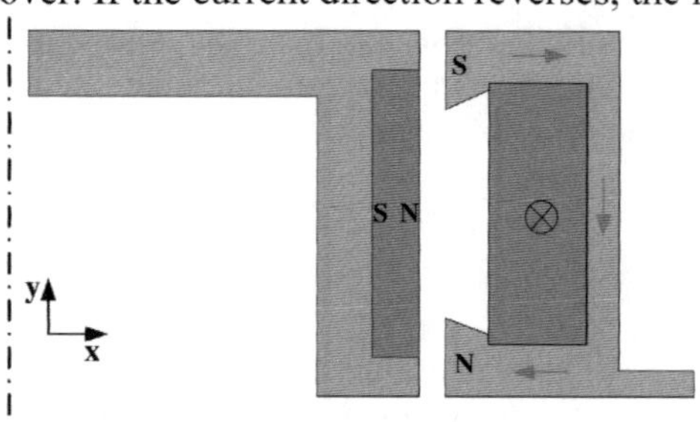

Figure 2. Magnetic configuration.

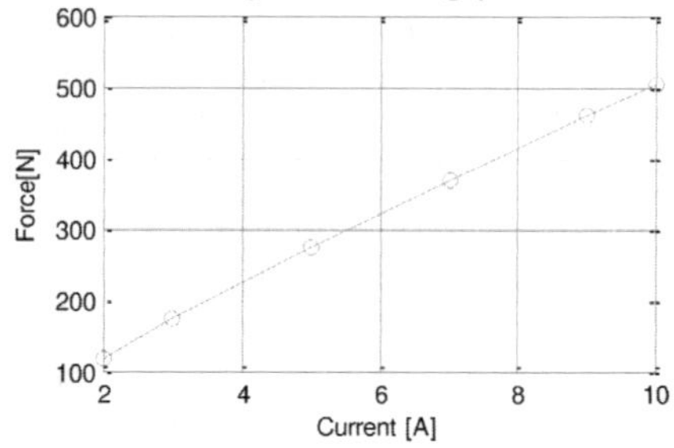

Figure 3. Force-current relation.

Denote the magnetic flux flowing through the air gaps between the upper magnetic poles, and between the lower magnetic poles as Φ1, Φ2 respectively. Then, for the magnetic configuration shown in Figure 2, we have:

$$\begin{aligned}\Phi_1 &= B_1 A = \bar{\Phi}_1 - \tilde{\Phi} \\ \Phi_2 &= B_2 A = \bar{\Phi}_2 + \tilde{\Phi}\end{aligned} \tag{1}$$

where B_1 and B_2 are the magnetic flux densities in the upper and lower air gaps, A is the area of the magnetic poles. $\bar{\Phi}_1$ and $\bar{\Phi}_2$ are the biasing flux generated by the permanent magnets, and $\tilde{\Phi}$ is the excitation flux generated by the coil.

According to [9], when the mover has very small displacements Δy from the central position, the variation of flux linkage is obtained as:

$$\begin{aligned}\Delta\psi &= N(\Delta\Phi_1 + \Delta\Phi_2) \\ &= 2NB_m d_g \,\Delta y\end{aligned} \tag{2}$$

and the thrust force can be calculated as:

$$F = I\frac{\Delta\psi}{\Delta y} = 2NB_m d_g \cdot I = K_i \cdot I \tag{3}$$

where N is the number of coil turns, B_m is the biasing flux density in the air gaps generated by the permanent magnet, d_g is the central diameter of the air gap, K_i is the magnetic force coefficient, and I is the amplitude of the coil current. It can be seen that the thrust force F generated by the electromagnetic actuator is proportional to the amplitude of the coil current I. The experimentally measured force -current relation is shown in Figure 3, which verifies the linearity of the actuator.

3. Experimental setup

In order to test the performance of the proposed novel hybrid vibration isolators, an experimental setup is built as shown in Figure 4. A shaker is used as the vibration source to simulate the vibration effects of rotary machines. Four hybrid vibration isolators are installed between the top and base steel plates at the four corners. Four BE85 rubber vibration isolators are installed underneath the base plate to support the whole experimental setup. An accelerometer is installed nearby the shaker to measure the reference vibration signals. The embedded accelerometers within the hybrid vibration isolators are used to measure the vibrations transmitted to the base. A multichannel signal conditioner is used to power the ICP accelerometers and transfer the acceleration signals to the dSPACE control system for active vibration suppression. Control algorithm is developed in a desktop computer with Matlab software, and is downloaded into the dSPACE control system for real time vibration control. Power amplifiers receive the command signals generated from the dSPACE control system, and power the shaker and hybrid vibration isolators.

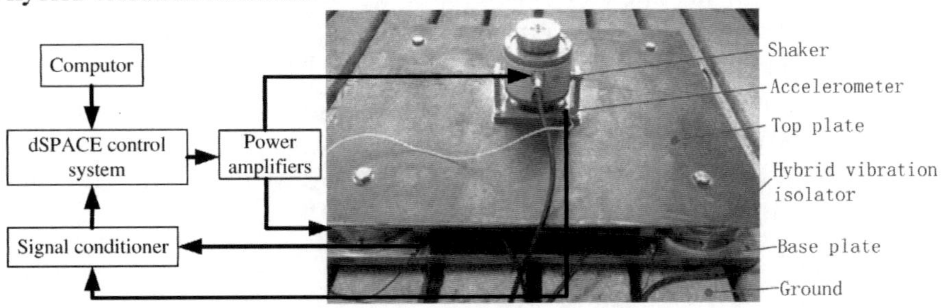

Figure 4. Experimental setup for hybrid vibration isolation tests.

4. Controller design

A single channel FxLMS control algorithm is developed in this work for prototype demonstration. Only the hybrid vibration isolator on the left-down corner of the experimental setup is controlled in this experiment. The block diagram of the control system is given in Figure 5. *P(z)* is the primary path dynamic response relation between the reference acceleration signal (*x(n)*) measured close to the shaker and the error acceleration signal (*d(n)*) measured nearby the hybrid vibration isolator. *S(z)* represents the secondary path frequency response function from the control signal (*y(n)*) to the error acceleration signal ($d_y(n)$) measured nearby the hybrid vibration isolator. The total error acceleration signal (*e(n)*) sensed by the accelerometer is $d(n)$-$d_y(n)$. The reference signal (*x(n)*) and error signal (*e(n)*) are passed to the FxLMS controller to calculate the control signal (*y(n)*).

In the FxLMS controller structure, a finite impulse response (FIR) filter (denoted as $\hat{S}(z)$) is used to model the secondary path (*S(z)*) of the control system, and the coefficients of the FIR filter ($\hat{S}(z)$) are identified offline using the least mean square (LMS) method. The estimated secondary path model ($\hat{S}(z)$) is used to filter the reference signal (*x(n)*). Another FIR filter (*W(z)*) is used as the controller, and its coefficients are updated in real time using LMS method based on the filtered reference signal (*r(n)*) and the error signal (*e(n)*). The controller coefficients and the control signal can be calculated with the following equations[10]:

$$\mathbf{w}(n+1) = \mathbf{w}(n) + \mu \mathrm{e}(n)\boldsymbol{r}(n) \tag{4}$$

$$\mathrm{y}(n) = \mathbf{w}(n)^{\mathrm{T}}\mathbf{x}(n) \tag{5}$$

where μ is the step size, $\mathbf{w}(n) = [\mathrm{w(n)\ w(n-1)\ ...\ w(n-M+1)}\ ]^{\mathrm{T}}$ is the controller coefficients vector, M is the tap length of the controller (*W(z)*), $\boldsymbol{r}(n) = [\mathrm{r(n)\ r(n-1)\ ...\ r(n-M+1)}\ ]^{\mathrm{T}}$, $\boldsymbol{x}(n) = [x(\mathrm{n})\ x(\mathrm{n}-1)\ ...\ x(\mathrm{n-M+1})\ ]^{\mathrm{T}}$.

5. Results analysis

During the vibration control experiments, the step size (μ) and tap length (M) of the controller (*W(z)*) are selected to be $\mu=10^{-5}$, M=500. A periodic disturbance which contains two different frequency components (55 Hz and 155 Hz) are generated with the shaker as the vibration source. The vibration level of the base plate at the hybrid vibration isolator location is measured and compared with and without control. When the controller is turned off, the base plate is vibrating severely, large peaks at 55 Hz and 155 Hz can be clearly seen from the frequency spectrums of the base plate vibration signals, as shown in Figure 6. When the controller is turned on, the acceleration amplitude of the base plate drops dramatically, the frequency spectrums of the base plate vibration signals at 55 Hz and 155 Hz are suppressed by 21.2dB and 27.9dB, respectively.

6. Conclusion

The design of a novel hybrid vibration isolator which contains a linear electromagnetic actuator and a rubber vibration isolator is presented in this paper. The actuator and the rubber vibration isolator are configured in a parallel structure with the rubber vibration isolator installed inside the mover of the actuator. The magnetic force generated by the actuator has a linear relationship with the amplitude of the coil current which is verified through experimental measurements. An experimental setup is built with the proposed hybrid vibration isolators, and a FxLMS controller is developed for prototype demonstration. The experimental results show that the vibration level of the base plate and the acceleration transmissibility are decreased significantly with the proposed hybrid vibration isolator and the FxLMS control algorithm.

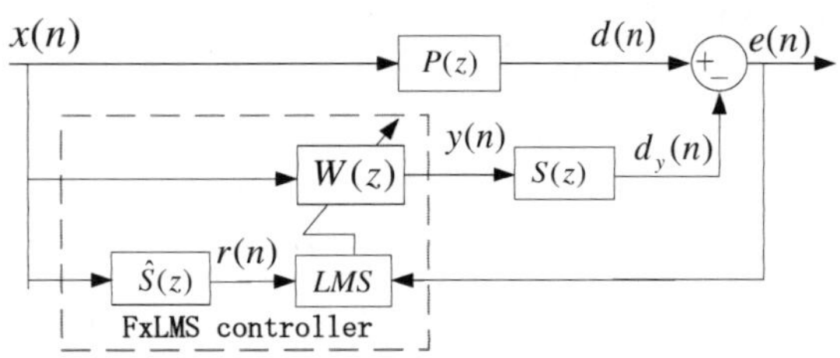

Figure 5. Block diagram of the control system using FxLMS algorithm.

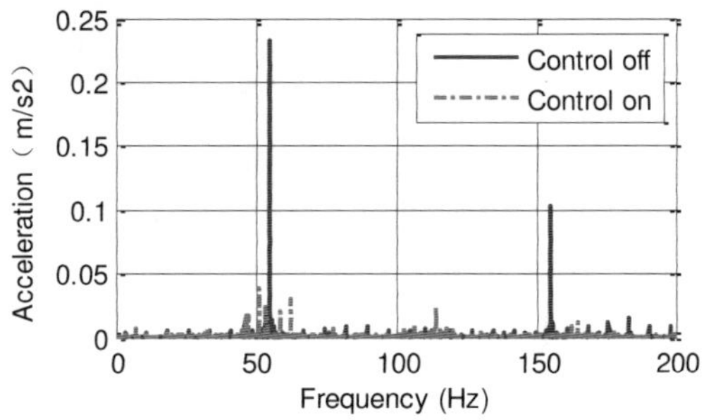

Figure 6. Frequency spectrums of the base plate vibration signals, with and without control.

References

[1] Chavan V S, Askhedkar R, and Sanap S B 2013 *Int. J. Eng. Res. Appl.* **3** 1423–28
[2] Chen P and Shih M, 2011 *J. Vib. Control* **17** 1325–36
[3] Gao J, Ji H and Qiu J 2015 *J. Vib. Shcock* **34** 141–8
[4] Chen P and Shih M 2007 *J. Vib. Control* **13** 1553–71
[5] Liu Y 2005 *J. Low Freq. Noise, Vib. Act. Control* **24** 181–90
[6] Ahn K G, Pahk H J, Jung M Y and Cho D W1996 *J. Sound Vib.* **192** 793–805
[7] Daley S, Johnson F A, Pearson J B and Dixon R 2004 *Control Eng. Pract.* **12** 465–74
[8] Li Y, He L, Shuai C and Lv Z 2014 *Tech. Acoust.* **33** 1–5
[9] Lu Q, Yu M, Ye Y, Fang Y and Zhu J 2011 *IEEE Trans. Magn.* **47** 4211–14
[10] Chang D and Chu F 2013 *IEEE Signal Process. Lett.* **20** 1122–25

Acknowledgement

This research was supported by the National Natural Science Foundation of China (Grant No. 51705175) and the China Postdoctoral Science Foundation (Grant No. 2017M612444).

IOP Conf. Series: Materials Science and Engineering **361** (2018) 012013 doi:10.1088/1757-899X/361/1/012013

Improving the performance of EDM deep hole using multi-hole interior flushing electrode

S. Chuvaree, K. Kanlayasiri

Department of Industrial Engineering, Faculty of Engineering, King Mongkut's Institute of Technology Ladkrabang, Bangkok 10520, Thailand

E-mail : c.suppawat@gmail.com

Abstract. In conventional flushing, when increasing the machining depth, the flushing ability becomes insufficient for ejection of the debris from machining gap. Debris concentration in the gap spark causes a process instability, low performance and high machining time etc. This paper proposes an innovative method of flushing technique for electrical discharge machining (EDM) of deep hole process. Tool electrode was specifically fabricated for flushing on the efficiency and stability of machining to promote the ejection of debris from machining gap. The performance of the deep hole process by electrical discharge machining under different flushing condition are presented. The results showed that the flushing through the tool electrode and multi-hole interior flushing are more preferred than side flushing. The result include 26.10% reduction in machining time (50 mm. in depth) when using multi-hole interior flushing electrode. In addition, the application of electrode multi-hole interior flushing electrode with electrode rotation improves the material removal rate (MRR) by 35.28% with 58.04% higher surface roughness (Ra).

1. Introduction

One of the important factors in a successful deep-hole EDM operation is the removal of debris from the machining gap. The flushing of these debris out of the machining gap is very important in preventing secondary spark and enhancing performance of machining. In the EDM process, the workpiece and tool electrode are submerged in dielectric fluid. The material removal process takes place as effect of electrical sparks occurring between tool electrode and workpiece that are separated by dielectric fluid [1]. The dielectric fluid is fed into small gap to remove debris particles that are generated by EDM and maintain the stability of machining condition. If the debris accumulates in the machining gap, the electrical conductivity of the dielectric liquid will be increased, leading to secondary spark phenomena as well as process instability and poor machining conditions. The influence of flushing of dielectric plays a crucial role during the whole process. Improper flushing can result in sparking that high electrode wear, increased machining time and surface finish [1–2].

To enhance the flushing of debris, the desired machining can be achieved by using a side jet flushing of fluid directed into the machining gap. However, the effectiveness of side flushing is limited when the machined depth is increased due to the ejection of debris from machining gap is extremely difficult and it produces a tapered opening in the workpiece [3–7]. L Li *et al.* [8], studied flushing technique using bundled electrode compared with a solid electrode in machining Ti6Al4V. It has been proved that bundled electrodes can endure a much higher peak current than solid electrode

Content from this work may be used under the terms of the Creative Commons Attribution 3.0 licence. Any further distribution of this work must maintain attribution to the author(s) and the title of the work, journal citation and DOI.
Published under licence by IOP Publishing Ltd

which results in a substantially higher material removal rate (MRR) and a comparably lower tool wear ratio (TWR). R Nastasi *et al.* [9], proposed the application of deep slotted tools in EDM drilling process compared with cylindrical and helical electrode. The results showed that at a machining depth of 5 cm, the higher MRR could be obtained using deep slotted tool.

In the EDM deep hole process, the viscous resistance in the narrow discharge gap causes difficulty in removal of debris and bubbles from the sparking area and abnormal discharge are occurred and resulting in extensive performance of deep hole process. Since, several methods such as electrode jump high, side flushing, mono hole inner flushing and reverse flow flushing have been established [3–6], but the application of the flushing technique for deep hole drilling has been very less reported.

This research proposes an EDM machining technique for deep hole by using multi-hole interior flushing electrode. The material removal rate (MRR), working time (Wt.) and surface roughness (Ra) have been carried out on the AISI P20 tool steel. The performance of the deep hole process by electrical discharge machining under different flushing condition are presented.

2. Experimental Setup

The experiments were carried out on 50 mm. depth of cylindrical hole in AISI P20 tool steel with ARISTECH CNC-EDM model 430. The EDM was carried out macro-scale machining with copper electrode as outside diameter of 12 mm, inner hole diameter 8.4 mm and the bottom of electrode was specifically designed that the dielectric running directly through internal diameter of tool electrode using multi channels inferior flushing (12 holes with a diameter of 1 mm.) as shown in Figure 1.

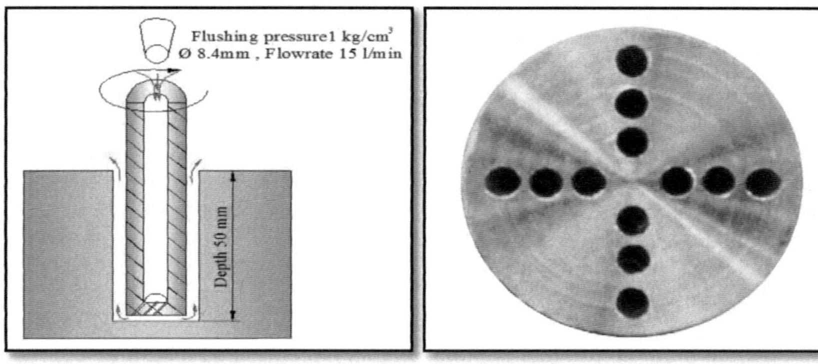

Figure 1. Tool electrode using multi holes inferior flushing

The special rotary chuck has been fabricated for EDM deep hole process and mounted to the EDM machine. The tool rotation helps in improving the flushing ability and removal of debris from the machining gap during EDM of deep hole process [10]. During the EDM deep hole process the flow rate of dielectric flushing was set on a constant value. The experimental conditions were shown in Table 1. The material removal rate (MRR) is defined as equation (1)

$$\text{MRR (g/mm}^3\text{)} = \text{(initial weight - final weight) / machining time} \quad (1)$$

Table 1. Experimental conditions

S.N.	In put parameters	Values
1	Pulse on-time	150 μs.
2	Pulse off-time	20 μs.
3	current	15 A.
4	Voltage	150 V.
5	Electrode rotational speed	0, 35, 70, 105, 140 and 175 rpm.
6	Polarity (Electrode)	+

7	Flushing Pressure	$1 kg/cm^2$
8	Machining depth	50 mm.
9	Dielectric fluid	EDM oil-TOTAL DIET MS 700

3. Result and Discussion

3.1 Machining Time

When increasing the machining depth, the flushing of debris out of the machining gap is very important problem. In this section, a comparison between solid electrode using side flushing and solid electrode using multi-hole interior flushing is observed. As the results shown in Figure 2, it is evident that the reduction of machining time resulted from the influence of flushing condition under multi hole inner flushing. When the machining depth is increased, the multi-hole interior flushing disposes the debris out of the machining gap. Consequently, the spark intensity becomes better than side flushing and thus the machining time is reduced. At a depth of 50 mm, machining time is reduced from 339.01 minutes to 250.53 minutes when using multi-hole interior flushing electrode.

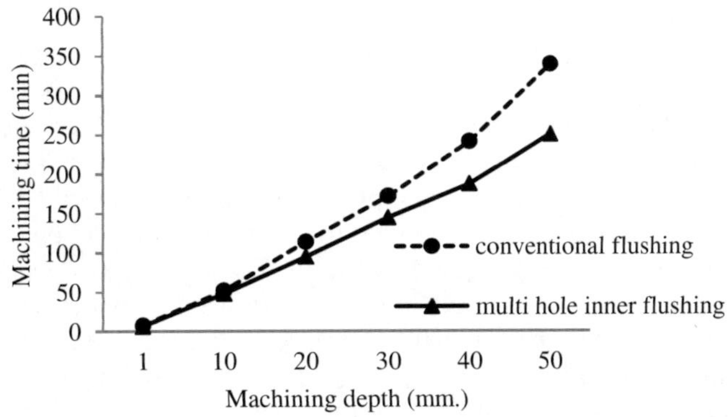

Figure 2. Machining time VS machining depth with different flushing condition

3.2 Material removal rate

In this experiments, the obtained MRR under conventional side flushing and multi holes inner flushing with different electrode rotational speed are illustrated in Figure 3. Under a multi hole inner flushing electrode with electrode rotation, the highest MRR of 22.485 mm^3/min. is achieved. It is assumed that the effect of electrode rotation couple with multi holes inner flushing causes the increase in MRR. The fresh dielectric is fed through multi-hole interior electrode and flowed out of small gap between electrode and workpiece to remove debris particles continuously. Moreover, the electrode rotation during EDM deep hole process promotes cleaning of debris in machining gap. As a result, complete spark occurs and is better than side flushing electrode.

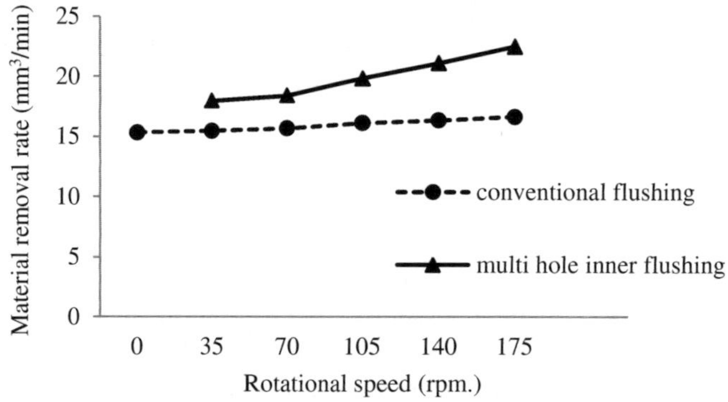

Figure. 3 MRR under conventional side flushing and multi hole inner flushing with different electrode rotational speed at a depth of 50 mm.

3.3 Surface Roughness

In this experiments, the effects of improved flushing increases MRR and reduces machining time. At the same time, process stability is increased because tool rotation induces fresh dielectric into the discharge gap as the used dielectric is forced out due to the centrifugal force. According to Figure 4, it is observed that when using conventional electrode via side flushing the surface roughness is improved with increasing electrode rotation speed. The surface roughness with the fixed electrode is 6.125 μm, whereas it is approximately 4.345 μm for the electrode rotation at 175 rpm. It is assumed that the debris is uniformly distributed and re-solidified on the surface leading to smoother surface. However, the flushing effect is limited when using multi-hole interior flushing electrode. Too much cleaning of the spark gap always increases surface roughness because of the rapid recovery of dielectric strength in spark gap. As a result, the spark intensity is very violent and thus higher MRR is obtained with a higher surface roughness [3, 5, 7]. Figure 5, shows the surface topography analysis by optical profilometer of the machined surface under different electrode flushing and electrode rotation.

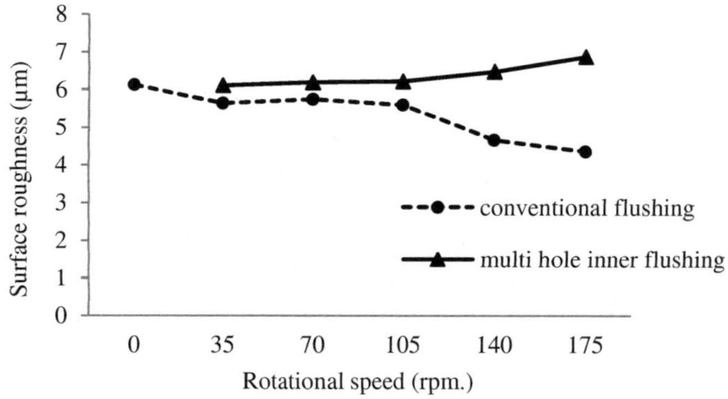

Figure 4. Influence of electrode rotation on surface roughness under conventional side flushing and multi hole inner

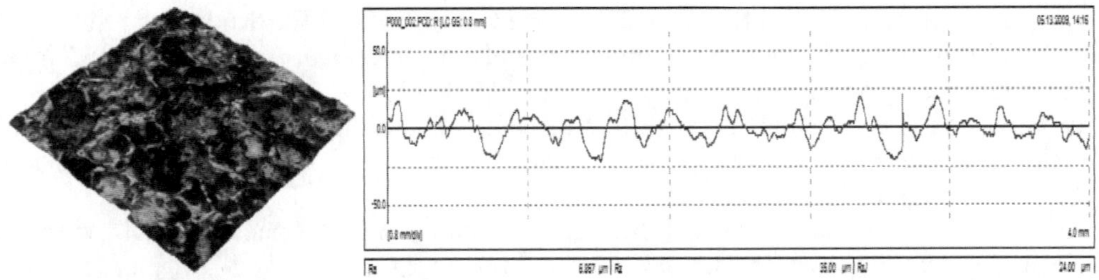

a) Multi-hole flushing electrode at 175 rpm of electrode rotation (Ra : 6.867 μm)

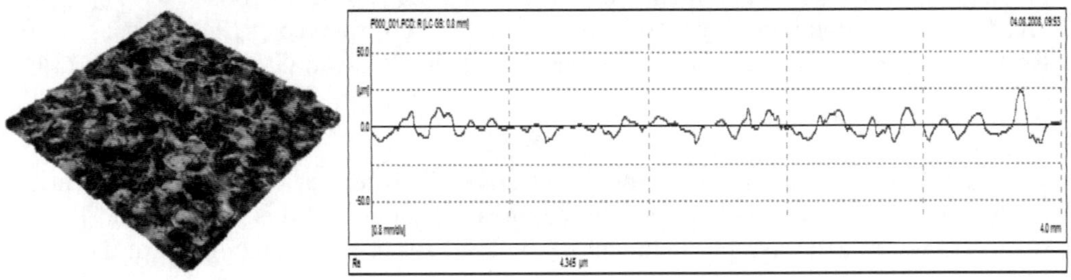

b) Side flushing electrode at 175 rpm of electrode rotation (Ra : 4.345 μm)

Figure 5. Surface topography under different electrode flushing and electrode rotation

4. Conclusion

In this present work, a new flushing method is proposed and the performance of EDM deep hole process using newly proposed multi-hole inferior flushing is compared with side flushing on the AISI P20 tool steel. The following conclusions can be drawn from the experiment.

- The improved flushing increases MRR and reduces machining time. At the same time, process stability increases because tool rotation induces fresh dielectric into discharge gap as the used dielectric is forced out due to the centrifugal force.
- 26.10% reduction in machining time (50 mm. in depth) when using multi-hole interior flushing electrode is achieved.
- The multi-hole interior flushing electrode with electrode rotation improves the material removal rate (MRR) by 35.28% with 58.04% higher surface roughness (Ra).

References

[1] El-hofy H A 2005 *Advanced Machining Processes : Nontraditional and Hybrid Machining Processes,* McGraw-Hill Mechanical Engineering Series.

[2] Jahan M P 2015 *Electrical discharge machining (EDM) : Types, Technologies and Applications*, Nova Science Publishers, Ins, New York.

[3] Munz M, Risto M and Haas R 2013. Specifics of Flushing in Electrical Discharge Drilling. *Procedia CIRP*. **6** pp 83−88

[4] Pichai J and Apiwat M 2011. A study of hole drilling on stainless steel AISI 431 by EDM using brass tube electrode. *International Transaction Journal of Engineering Management & Applied Sciences & Technologies*. Vol **2(4)** pp 471−481

[5] Yilmaz O and Okka M A 2010. Effect of single and multi-channel electrodes application on EDM fast hole drilling performance. *The International Journal of Advanced Manufacturing Technology*. **51(1)** pp 185−194

[6] Skoczypiec S, Machno M and Bizon W 2015. The Capabilities of Electrodischarge Microdrilling of High Aspect Ratio Holes in Ceramic Materials. *Management and Production Engineering Review.* Vol **6(3)** pp 61−69

[7] Cetin S, Okada A and Uno Y 2004. Effect of Debris Distribution on Wall Concavity in Deep-Hole EDM. *JSME International Journal Series C Mechanical Systems, Machine Elements and Manufacturing.* **47(2)** pp 553−559

[8] Li L, Gu L, Xi X and Zhao W 2012. Influence of flushing on performance of EDM with bunched electrode. *The International Journal of Advanced Manufacturing Technology.* **58(1)** pp 187−194

[9] Nastasi R and Koshy P 2014. Analysis and performance of slotted tools in electrical discharge drilling. *CIRP Annals - Manufacturing Technology.* **63(1)** pp 205−208

[10] Dwivedi A P and Choudhury S K 2016. Increasing the Performance of EDM Process Using Tool Rotation Methodology for Machining AISI D3 Steel. *Procedia CIRP.* **46** pp 131−134

Acknowledgment

The authors would like to thank Department of Industrial Engineering, Faculty of Engineering, King Mongkut's Institute of Technology Ladkrabang for financial support of this research and Rajamangala University of Krungthep and Olympus (Thailand) Co., LTD for providing facilities for the experiment.

IOP Conf. Series: Materials Science and Engineering **361** (2018) 012014 doi:10.1088/1757-899X/361/1/012014

Method of load calculation of electrical drives of rolling mills during heavy plate manufacturing

V R Gasiyarov[1], A A Radionov[1], A S Karandaev[1], V R Khramshin[2]

[1]South Ural State University, 76, Lenin Avenue, Chelyabinsk 454080, The Russian Federation

[2]Nosov Magnitogorsk State Technical University, 38, Lenin Av., Magnitogorsk, 455000, The Russian Federation

E-mail: gasiyarovvr@gmail.com

Abstract. During manufacturing of heavy strips and plates from difficult-to-form steel grades on sheet and wide-strip rolling mills there is a problem of exact calculation of power parameters of rolling. It is pointed out that known methods do not provide required precise information at high deformation zones in a stand. An improved calculation method for rolling pressure and load torque of electrical drive engine is suggested further. This method considers peculiar strain of product in high deformation zones and its accelerated motion in reverse mode. Improved analytical dependences for rolling pressure and load torque of electrical drive are reviewed. Control flow chart and main screen of developed software for calculation of wide-strip rolling parameters are presented. Comparative analysis of calculation data and experimental results of engine equivalent currents during tube steels rolling is made. As a result improved precision of calculation method is justified.

1. Introduction

Service experiment for plate mill 5000 and mastering rolling of hollow billet on continuous wide-strip mill 2000 predetermine the need of theoretical analysis of power parameters of heavy plate rolling [1,2]. Speed and load charts of electrical drives of a rolling mill are interdependent constraining factors applied to technological process. This is why their analysis is required for development of new and improvement of available technological software.

Process of rolling of a heavy plate on continuous hot mills and on reversible sheet mills is different. Rolling on continuous hot mill is characterized by variable speed. Front of the piece is collected in sequence by each stand and a baller on low threading speed. Then all stands, a rolling table and a baller speed up simultaneously to maximal operating speed. Lowing down of stands is individual after a trailing end leaves a certain stand.

Drafting on reversing mill 5000 is performed by one horizontal four-high stand and a two-roll edger stand [3]. Universal stand operates in reverse mode – slabbing is performed within several passes. Before and after a stand there are turn-around machines with side guides to ensure 90-degree turn of sheet slabs [4]. Acceleration and speed reduction of electrical drive are carried out with plates in rolls. Acceleration takes place immediately after the plate is collected by a roll; speed reduction torque is determined by preselected speed of plate release on an outgoing table.

Content from this work may be used under the terms of the Creative Commons Attribution 3.0 licence. Any further distribution of this work must maintain attribution to the author(s) and the title of the work, journal citation and DOI.
Published under licence by IOP Publishing Ltd

Analysis of loading modes of rolling mill electrical drives is a time consuming task associated with numerous mathematical calculations. Specialized software significantly facilitates calculations. To implement computerized analysis of operation modes of engines of stand main electrical drives we need to develop new methods and software.

2. Set up of the problem

In practice, to calculate power parameters of rolling one traditionally applies methods developed based on the works of academy fellow A.I. Tselikov. In most cases, the results for hot rolling of plates are reasonable for engineering calculation. The author of the publication [5] examined methods of numerical definition of rolling pressure. To solve the differential equation

$$\frac{d\sigma_x}{dx} - \frac{P_x - \sigma_x}{y} \cdot \frac{dy}{dx} - \frac{t_x}{y} = 0 \tag{1}$$

unit roll pressure P_x -stress σ_x dependence was analyzed. Main vertical intensity formula was used

$$\sigma_1 = \left(P_x \cdot \frac{dx}{\cos\phi} \cdot \cos\phi \pm t_x \cdot \frac{dx}{\cos\phi} \cdot \sin\phi \right) \cdot \frac{1}{dx} \tag{2}$$

The second term, due to its small amount in comparison with the first member, was not considered by the author. In this case problem solution (1) provides the following formula for rolling pressure P calculation

$$P = \frac{b_0 + b_1}{2} \cdot \frac{l}{\Delta h} \cdot \frac{k \cdot h_\gamma}{\delta} \cdot \left(\left(\frac{h_0}{h_\gamma} \right)^{\delta} + \left(\frac{h_\gamma}{h_1} \right)^{\delta} - 2 \right) \frac{\partial^2 \Omega}{\partial u^2} \tag{3}$$

In equations (1) – (3) the following keys are introduced: x, y – coordinates of reference section of deformation zone on roll face; φ – angle between the tangent to roll bow at calculation point and level; t_x – specific friction force between rolling metal and roll face; h_0, h_1 and b_0, b_1 – thickness and width of a sheet slab before and after rolling; Δh – draught; l – length of deformation belt; h_γ – plate height in neutral plane; δ –factor depending on roll parameters.

For lower deformation zones, which occur during light plate rolling, the given equations provide reasonably precise calculations of power parameters. However, during heavy plate rolling it is unacceptable to dismiss the second term of equation (2). Given that rolled sheet slab is 350 mm thick and working roll diameter is 1210 mm, angle φ is 20°, thus, the second term of equation is relevant.

As a result we face the need to develop an improved calculation method of rolling pressure and rolling torque considering the effect of thickness factor. Moreover, plates move unevenly on reversing mill because they leave rolls with acceleration. That is why we must take into account inertial force U of metal (figure 1) [5]. When developing a new method it is important to consider that the ratio between level of deformation zone and radius of working rolls significantly influence static rolling torque.

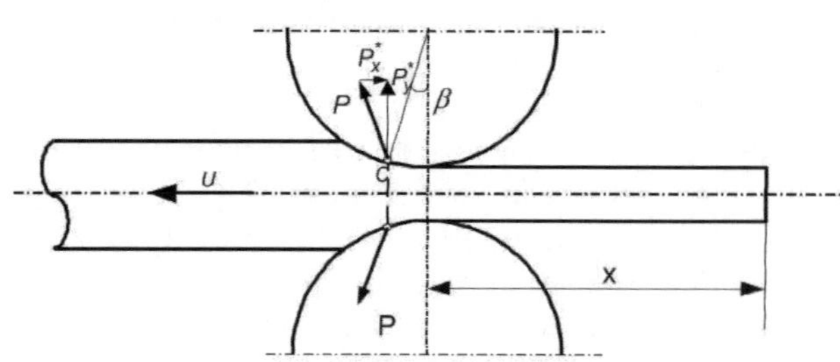

Figure 1. Direction of forces acting on rolls during accelerated movement of rolled metaFl.

3. Main part

What makes the new method distinctive are the more precise equations used for calculation of rolling pressure and static torque during heavy plate rolling. Torque equation is written as

$$M_r = 2 \cdot P \cdot \psi \cdot \sqrt{R_r \cdot (h_0 - h_1) - \frac{(h_0 - h)^2}{4}} \tag{4}$$

where ψ is a position coefficient of thrust force acting on rolls; R_r is working roll radius.

Rolling pressure is calculated based on dependences

$$P = \sqrt{P_x^{*2} + P_y^{*2}};\ P_x^* = \frac{1}{2} \cdot \frac{G}{g \cdot l_1} \cdot (l_1 - l_x) \cdot \frac{j}{\lambda} + \frac{G}{g \cdot l_1} \cdot l_x \cdot j;\ P_y^* = \frac{b_0 + b_1}{2} \int_0^{\alpha} P_\phi;$$

$$P_\phi = \mp t_x \cdot tg\phi \mp \frac{8 \cdot R_r^2}{(h + 2 \cdot R_r)\sqrt{h \cdot (h + 4 \cdot R_r)}} \cdot arctg\left(\frac{h + 4 \cdot R_r}{\sqrt{h \cdot (h + 4 \cdot R_r)}} \cdot tg\left(\frac{\phi}{2}\right)\right) \mp \tag{5}$$

$$\mp \frac{2 \cdot R_r}{h + 2 \cdot R_r} \cdot \ln \left| \frac{tg(\phi/2) + 1}{tg(\phi/2) - 1} \right| + k \cdot \ln \left| \frac{h}{2} + R_r \cdot (1 - \cos\phi) \right| + k \cdot \left[1 - \ln \frac{h}{2} \right]$$

where G – weight of the plate; L_l – length of the plate at delivery; x – rolling out as a result of plate travel; l_x – length of plate delivered from the rolls at a given torque; j – plate acceleration at delivery; g – free fall acceleration. The remaining parameters are noted above.

The author of the publication [6] has deducted a ratio for calculation of position coefficient of thrust force acting on rolls

$$\psi = \frac{a_M}{l} = \frac{1}{3} \cdot \frac{1 + 2 \cdot p_0/p_1}{1 + p_0/p_1} \tag{6}$$

Analysis of equation (6) leads to conclusion that at $l/h = 1$ and $p_0/p_1 \to \infty$ highest value $\psi = a_M/l$ is 0.67. At $l/h \to 0$ value of ψ comes close to 0.5, and at $l/h \to \infty$ – to 0.33. Analytical angle value β for different distribution laws for surface stress was calculated by well-known Russian scientists metallurgists A.A. Korolev [7], N.S. Spiridonov [8], N.M. Kirilin [9] and others. The calculations were made for those cases when frictional forces were in proportion to unit roll pressure. Similar research was conducted to identify the conditions for frictional force consistency as described in the publication [10]. When calculating the arm, the author [9] assumed that unit rolling pressure is distributed according to equations by Tselikov [11]. As a result of his research he concluded that with increase of drafting, the relation between angle β to entering angle α (arm coefficient ψ) decreases from 0.5, at light draft, to 0.4...0.45 at approximately 0.6 draft.

The results of researches by E.S. Rokotyan [12] and G. Valkvist [13], conducted in working environment on blooming and sheet mills are of our main interest. Analysis of their research results shows that value of angles β and α ratio during hot rolling ranges from 0.30...0.35 to 0.55...0.57. At that, low values correspond with lighter plates (2.5-5 mm) and upper values correspond with heavier plates (10-20 mm). These values rise when rolling temperature increases and draft decreases. For one thickness value variances ψ are relatively insignificant and do not exceed 15...20%. However, all these researches do not precisely explain how the angle β will change during heavy plate rolling.

To find how the angle of application point of rolling pressure resultant force changes during heavy plate rolling we conducted a series of experiments on rolling mill 5000. Steel grades K52, K56, 09G2S, 15HSND, 10G2FBU, 15ps, X70, PC D40 were used. Change in angle of application point of rolling pressure resultant force in relation to entering angle is presented on figure 2. The analysis shows that angle β for each steel grade relates in the same way to angle α, so it is possible to conclude that it does not depend on properties of steel, but rather on level of deformation zone.

IOP Conf. Series: Materials Science and Engineering **361** (2018) 012014 doi:10.1088/1757-899X/361/1/012014

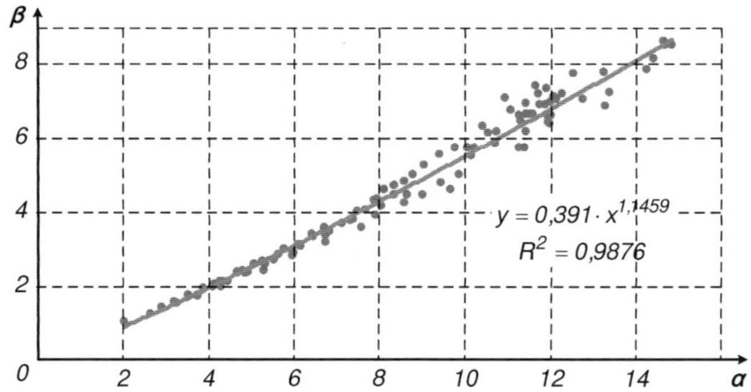

Figure 2. Ratio $\beta = f(\alpha)$ for rolling on mill 5000.

General tendency of change of angle β on figure 2 is reflected by power characteristic. Dependability approximant R^2 demonstrates how many points from presented selection are contained on the curve (the higher R^2, the better representation of dynamic pattern of approximant parameter). In the given case dependability is 98%, approximate connection $\beta = f(\alpha)$ is written as:

$$\beta = 0{,}391 \cdot \alpha^{1{,}1459} \tag{7}$$

A theoretical model for calculation of power parameters of rolling was developed based on equations (4)-(6). Theoretical model is actualized as software components in graphic environment of simulation study Simulink in MATLAB Packet.

Program flow and interface

We developed program flow and specialized software for computerized analysis of power parameters of sheets and heavy plates rolling. Flow chart of main software is shown on figure 3. Software is written in Object Pascal in visual object oriented programming environment for creation of applications for Windows Borland Delphi 7.0. This environment was chosen due to its simple terms for creation of applications and a rather rich library of generic components. Variety of basic primary objects have allowed for coding a program using standard libraries, and, thanks to flexibility of each component's settings it was possible to create the program in a unified style.

A compiled program has one file with the specified name. Figure 4 represents a window of the program made for calculation of rolling patterns of heavy strips in roughing and finishing mill groups of the mill 2000. There is also a version of the same program used for same calculations for the mill 5000.

The user window is divided into four sections. The top section contains general view of roughing and finishing mill groups. The section underneath the top one contains chart bars illustrating load on engines (percentage of nominal values). Bars are colored, depending on the load; green color of bars indicates load between 0% and 80%, yellow color – 80-90%, and red color – load exceeding 90%. The third section contains input data for calculation, including roll radius, gearbox ratios, nominal current and engine speed values, as well as rolling speed and acceleration. Finally, fourth section is intended for input of values of thickness, width and rolling temperature for each interstand space.

Operations with the program begin with data input (constants). They may be put in manually or imported from a preformed file. After having put in data, user initiates calculation. Input and output data may be shown on the display or printed out. Calculation logs as well as input and output data are exported in .xls file.

4. Theoretical and experimental study

To check efficacy of calculation method for operating modes as implemented in the software we applied it to calculate electromagnetic torque of 4-12 stand electrical drives of rolling mill 2000.

IOP Conf. Series: Materials Science and Engineering **361** (2018) 012014 doi:10.1088/1757-899X/361/1/012014

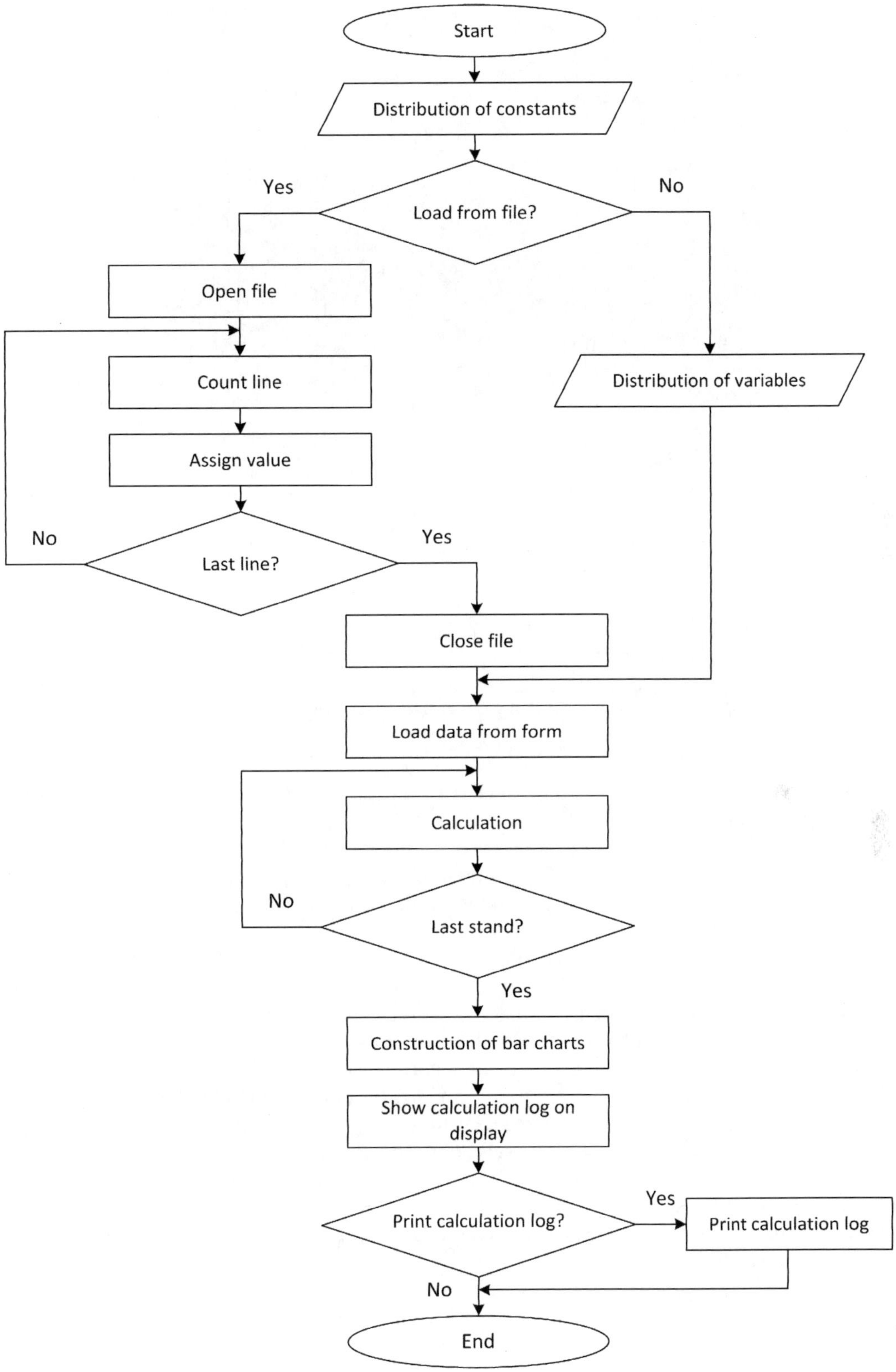

Figure 3. Control flow chart.

IOP Conf. Series: Materials Science and Engineering **361** (2018) 012014 doi:10.1088/1757-899X/361/1/012014

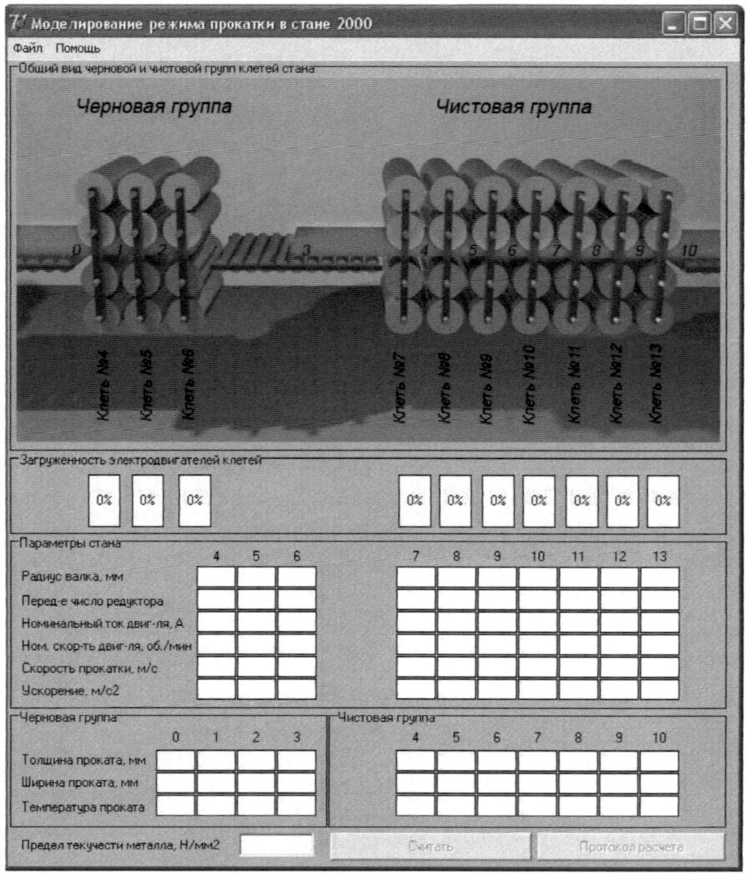

Figure 4. View of user window. **Legend (left to right, top to bottom):** roughing mill group, finishing mill group, stands, percentage of load of stands' engines, mill parameters, roll radius (in mm), gearbox ratio, nominal engine current (in A), nominal engine speed (in rpm), rolling speed (in m/sec), rolling acceleration (in m/sec^2), roughing mill group, rolling thickness (in mm), rolling width (in mm), rolling temperature, yield point of metal (in N/mm^2), finishing mill group.

The calculation was made for hollow billet rolling, steel grades 22GU, 09G2D, 09G2S, 09GSF, 13G1S-U. Steel grades and rolling modes correspond with experimental studies given in [14]. Results of efficacy estimation show that experimental and estimated data vary within the range from -7.8% to +6.6%. So the precision of calculations based on suggested method is satisfactory.

This method was applied to calculate equivalent current of direct current motors of electrical drives in rolling mill 2000 and electromagnetic torque of synchronous motor of stand rolls in rolling mill 5000. Results of calculations and experimental studies of operating modes of the given electrical drives are provided in table 1 and 2 respectively. For rolling mill 2000 operating modes are shown for hollow billet rolling using the given grades of steel, for rolling mill 5000 operating modes are for sheet slabs of the following steel grades: 10HSND, K56, K60, K65 и St3sp. Comparison of the results of theoretical and experimental study leads to conclusion that calculations are satisfactory: calculation error does not exceed 9-10%.

Table 1. Results of calculations and experimental studies of equivalent current of electrical drives in rolling mill 2000 during tube steel rolling.

Source of data	Equivalent current of engines by stands, A								
	№4	№5	№6	№7	№8	№9	№10	№11	№12
	steel grade 22GU, sheet slab size 5900x1250x250								
experiment	10624	9643	8564	7547	6421	8572	6058	6049	5552
calculation	10321	8974	8573	7904	6989	8162	6511	6060	5023
	steel grade 09G2D, sheet slab size 5550×1250×250								
experiment	11110	11231	12591	7808	6019	6750	5376	5691	7095
calculation	11056	11929	13434	8450	5588	6116	5768	5607	7015
	steel grade 09G2S, sheet slab size 9700×1650×250								
experiment	17418	15777	16464	7864	7865	9695	7623	8009	7801
calculation	17604	16770	16922	7547	7387	10458	7816	8342	8031
	steel grade 09GSF, sheet slab size 9300×1730×250								
experiment	16676	14420	15617	13028	11066	13744	11015	10570	8610
calculation	15096	13135	17144	12042	10488	12383	11838	9619	8987
	steel grade 13G1S-U, sheet slab size 6000×1730×250								
experiment	18356	16806	14947	11736	9685	12351	10490	11409	5736
calculation	17071	17634	15987	12519	9700	12084	10748	11352	5383

Table 2. Results of calculations and experimental studies of rolling torque in rolling mill 5000.

Steel grade	Sheet slab size, mm			Draft,	Rolling torque, kN	
	thickness	width	length	%	experiment	calculation
10HSND	300	3040	2400	6.08	2984	3284
K56	300	3120	2050	11.28	4195	4700
K60	300	2600	2400	9.89	4124	4522
K65	300	2660	2400	11.89	5761	6242
St3sp	300	2900	2700	9.67	3084	3318

Results of experimental rolling of sheet slab, 300 mm, different steel grades, on rolling mill 2000 and results of theoretical calculations based on traditional method and software-implemented method are given in table 3.

Table 3. Results of calculations and experimental studies of sheet slab 300 mm rolling.

Steel grade, sheet slab size	Experiment, calculation	Equivalent current of engines by stands, A/(% to experiment)						
		№7	№8	№9	№10	№11	№12	№13
steel grade 20, 8000×1750×300	experiment	12.5	12.6	12.8	12.2	12.0	8.9	-
	traditional method	13.4 +7.2%	12.1 -4.0%	12.0 -6.3%	11.9 -2.5%	10.5 -15.0%	4.8 -54%	-
	suggested method	12.9 +3.2%	12.8 +1.5%	13.0 +1.5%	12.7 +4.0%	12.8 +6.6%	8.6 -3.4%	-
08 ps, 8000×1330×300	experiment	9.9	8.4	10.2	8.7	8.9	5.6	5.4
	traditional method	10.7 +8.0%	9.0 +7.1%	11.0 +7.8%	8.9 +2.2%	10.2 +14.6%	6.1 +8.9%	4.1 -25%
	suggested method	10.3 +4.0%	9.0 +7.1%	10.6 +3.9%	9.1 +4.5%	9.0 +1.1%	5.9 +5.3%	5.0 -7.5%

Comparative analysis of the given data leads to conclusion that suggested method of calculation of power parameters during heavy sheet slab rolling provides more accurate results than traditional method. So, for the given range the error against experimental data for suggested method is within -7.5%...+7.1% range, while for traditional method it is within -15%...+14.6% range (analysis does not include calculation results of the last stands).

5. Conclusion

Experimental studies of operating modes of heavy plate mill 5000 electrical drives and wide-strip rolling mill 2000 have shown that direct application of simplified method by A.I. Tselikov is unacceptable when calculating power parameters of rolling of heavy plates from difficult-to-form steels. Calculating error for first passes on rolling mill 5000 is as high as 38%. It predetermined the need to develop a new more precise method.

Analytical dependences for rolling pressure during rolling of sheet slabs thicker than 250 mm are given. When developing a new method, particularity of rolling mill accelerated movement was considered. Angle of application point of rolling pressure resultant force in rolling mill 5000 stand was determined experimentally. It was found that this angle is defined by deformation zone configuration and does not depend on steel grade.

Program flow and software for computerized analysis of power parameters of rolling and operating modes of electrical drives in heavy plate and wide-strip rolling mill was developed. It was confirmed that the software provides calculations for rolling pressure and torque with error not higher than 7.5% for all passes.

Experiment has proved efficacy of calculated data, which lets us recommend the suggest method and software for extended application.

References

[1] Radionov A A, Karandaev A S, Khramshin V R, Andryushin I Yu and Gostev A N 2014 Speed and Load Modes of Rolling Hollow Billet at the Wide-Strip Rolling Mill *Proc of 2014 Int Conf on Mechanical Engineering, Automation and Control Systems (MEACS)* p 5 DOI 10.1109/MEACS.2014.6986841

[2] Khramshin V R, Evdokimov A S, Kornilov G P, Radionov A A and Karandaev A S 2015 System for Speed Mode Control of the Electric Drives of the Continuous Train of the Hot-Rolling Mill *Proc of the 2015 Int Siberian Conf on Control and Communications (SIBCON)* DOI 10.1109/SIBCON.2015.7147264

[3] Khramshin V R, Karandaev A S, Evdokimov S A, Andryushin I Yu, Shubin A G and Gostev A N 2015 Reduction of the Dynamic Loads in the Universal Stands of a Rolling Mill *Metallurgist* vol 59 **3-4** pp 315–23 DOI 10.1007/s11015-015-0103-8

[4] Khramshin V R, Khramshina E A, Karandaev A S, Gasiyarov V R and Voronin S S 2017 Control Methods and Systems Providing Reduced Consumption Index at Rolled Product Manufacture at Plate Mill *Proc of the IEEE NW Russia Young Researchers in Electrical and Electronic Engineering Conf (EIConRus)* pp 1540–4 DOI 10.1109/EIConRus.2017.7910865

[5] Tselikov A I 1946 Rolling mills (Moscow: Metallurgizdat)

[6] Weber K H 1959 Neeue Hütte **10** p 620

[7] Korolev A A 1952 Distribution of contact frictional forces in deformation zone Steel **8**

[8] Spiridonov N S 1956 Grapho-analytical method of finding the arm of full resultant force *Rolling production: proceedings Institute of Ferrous Metallurgy* **X**

[9] Kirilin N M 1959 Analytical determination of position of thrust force acting on rolls in simple rolling process *News of higher educational institutions mechanical engineering* 11 pp 28–32

[10] Sims R B 1956 Devy United Engineering vol 1 **2**

[11] Tselikov A I 1962 *Theory of forces analysis in rolling mills* (Moscow: Metallurgizdat)

[12] Rokotyan E S 1956 *Blooming mill performance improvement by means of calculation of strength and power capability of equipment* (Moscow: Mashgiz)

[13] Valkvist G 1957 *Research of power parameters in hot rolling* (Moscow: Metallurgizdat)

[14] Radionov A A, Andrjushin I Yu, Galkin V V and Gostev A N 2011 Minimum speed limit of 2000 mill electric drives during tube shell rolling *Bulletin of Nosov MSTU* **3** pp 20–2

Acknowledgments

IOP Conf. Series: Materials Science and Engineering **361** (2018) 012014 doi:10.1088/1757-899X/361/1/012014

This research was performed with financial support in the form of a grant from the President of the Russian Federation (Grant No. MD-979.2017.8).

IOP Conf. Series: Materials Science and Engineering **361** (2018) 012015 doi:10.1088/1757-899X/361/1/012015

Experimental Study on Failure Mechanism of Advanced High Strength Steels in Air Bending Process

N Pornputsiri and K Kanlayasiri

Department of Industrial Engineering, Faculty of Engineering, King Mongkut's Institute of Technology Ladkrabang, Bangkok 10520, Thailand.

E-mail : natthasak_idt@hotmail.com

Abstract. This research investigated the failure phenomena and underlying damage mechanisms in small curvature bending and spring-back angles of two advanced high strength steel (AHSS) sheets (TRIP780 and DP980). The study also examined the effect of the bending angle on the failure mechanism. In the analysis, this research utilized an optimal microscope (OM), scanning electron microscope (SEM) and energy-dispersive X-ray spectroscope (EDX). The research findings indicated that the bending force and the bending angle were positively correlated and that the spring-back and bending angles were inversely correlated. On the failure mechanism, at the 45° bending angle, the initial crack was identified on the outer region of the LD-rolling DP980 sheet, while no visible crack emerged for the TD-rolling TRIP780 until arriving at the 60° bending angle, indicating that, by comparison, the former possessed the lowest bendability and the latter the highest.

1. Introduction

The structural lightweight, energy saving and safety requirements of the automotive industry contribute to the search for steels with high strength and good formability. In response, numerous attempts have been made and one viable solution is advanced high strength steels (AHSS). The enhanced strength however affects the formability of the steel into complex-shape parts. The AHSS are typically of multiphase steels whose microstructures comprise diverse phases, e.g. the dual phase (DP) and transformation induced plasticity (TRIP) steels [1]. To ascertain the industrial applicability of the DP and TRIP steels, it is necessary that the effects of their microstructures on the bendability and failure mechanisms be thoroughly investigated. The bendability of metallic sheets refers to the limit of the forming process accomplished without failure. The bending limit could be defined as the amount of elongation on the outer surface of the bending zone, from which a minimum bending radius is deduced [2]. Furthermore, the limit is subject to a variety of parameters, including the ratio of sheet thickness to tool radius, material properties, surface and microstructure of the metal sheet.

It is well known that the fabrication history has a significant effect on bendability and must be included failure mechanisms in order to achieve accurate predictions of forming limits. Nakataki et al [3] investigated the bendability of dual-phase AHSS metal sheets and it was documented that the microstructure of these steels could be refined and homogenized for the bendability improvement. Kaupper and Merklein [4] proposed the failure phenomena and underlying damage mechanisms associated with the small curvature bending of AHSS using a novel comprehensive optical strain-based characterization technique. R'eche et al [5] studied the failure mechanism in air-bending of

Content from this work may be used under the terms of the Creative Commons Attribution 3.0 licence. Any further distribution of this work must maintain attribution to the author(s) and the title of the work, journal citation and DOI.
Published under licence by IOP Publishing Ltd

TRIP steel sheet and the results showed that the initial cracks occurred on the outer bent surface (roof) of the metal sheet. Despite the ubiquity and severity of metal failure in the industrial production, prior research has largely focused on the relationship between the bending angle and the damage mechanism at the expense of several other contributing factors of the metal failure. In other words, there exists no research that attempts to explain the relationship of the metal microstructure, rolling direction and bending angle to the bendability and the failure mechanism of the metal sheet has ever been carried out.

This current research has investigated the relationships between the bending angle and the bending force; and between the bending and spring-back angles of the TRIP780 and DP980 AHSS steel sheets under the traverse (TD) and longitudinal (LD) rolling directions. The effect of the bending angle on the failure mechanism was also determined. In addition, an optimal microscope (OM), scanning electron microscope (SEM) and energy-dispersive X-ray spectroscope (EDX) were deployed in the analysis.

2. Materials and experimental procedure

2.1. Materials

This research experimented with two types of AHSS sheets of identical thickness (1mm): the transformation induced plasticity (TRIP780) and dual phase (DP980) steels. Table 1 tabulates their tensile mechanical properties determined using the uniaxial tensile test at room temperature.

Table 1. Mechanical property at room temperature.

Material	Yield Stress (MPa)	Ultimate Tensile Stress (MPa)	Total Elongation (%)
TRIP780	589	983	24.7
DP980	638	1017	19

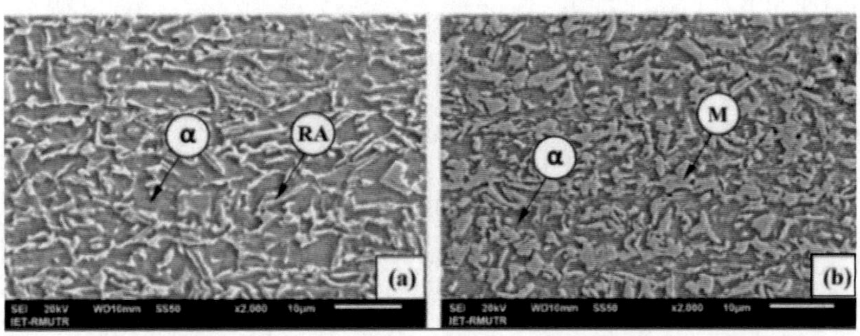

Figure 1. Initial microstructure of investigated steels (a) TRIP780 and (b) DP980.

In this research, the microstructure and chemical composition were respectively analyzed using SEM. Figure 1(a) illustrates the TRIP780 microstructure comprising the island-like retained austenite (RA) phases embedded in the ferrite matrix (α). Meanwhile, Figure 1(b) depicts the DP980 microstructure consisting of ferrite matrix (α) in which the martensitic phase (M) is dispersed in the form of islands.

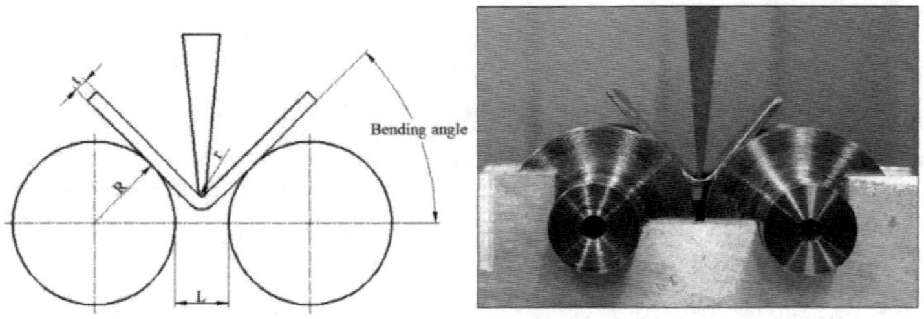

Figure 2. The schematics of the pure-bending experimental setup.

2.2. Experimental procedure

Figure 2 illustrates the pure-bending plane-strain experimental setup to investigate the failure mechanisms of the TRIP780 and DP980 steel sheets using the air bending test as per the ISO 7438:2010 standard. The punch (r) and roller (R) radii were respectively 0.25mm and 10mm. The metal sheets were 12mm in width (b), 30mm in length (l) and 1.0mm in thickness (t).

In the operation, the punch pressed downward while the support set remained stationary. The experiments were carried out by varying the bending angles in a 15° increment from 0° to 75° under the room temperature and without lubrication condition. The post-bending failure mechanisms were determined using photography technology by which the outer surface bent area was first investigated using a Digital Single Lens Reflex (DSLR) camera (CANNON, EOS 80D EF-S) and a Light Optical Microscope (LOM) (JENCO, GL7-290) to identify the initial crack. The specimens were then polished and etched prior to undergoing the macro-scale microstructure inspection using a 50X Optical Microscope (OM) (OLYMPUS, BX 60 M) to investigate the cross-sectional crack and crack propagation. The cross-sectional micro-scale failure mechanisms and the chemical composition at the bent area (or fracture surface) were further determined using SEM (JEOL, JSM-6510 LV) and EDX (OXFORD, X-MAX IE-350).

3. Experimental Results and Discussion

3.1. Relationships between the bending angle and spring-back angle

This part investigates the effects of the bending angle on the bendability of the TRIP780 and DP980 experimental metal sheets, given the two rolling directions (i.e. TD and LD).

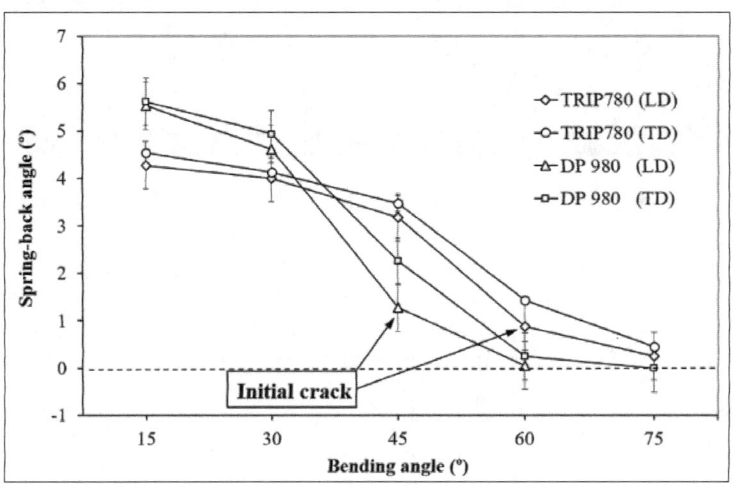

Figure 3. Relationships between the spring-back angle and bending angle of the DP and TRIP steels in the longitudinal and transverse rolling directions.

Figure 3 depicts the spring-back angles relative to the various bending angles (0°-75°) and the rolling directions. The experimental results showed that the spring-back angles were inversely correlated with the bending angles, whereby the spring-back angles decreased with increase in the bending angles. By comparison, the spring-back angles relative to the bending angles were greatest for the TD-rolling TRIP780 metal sheet, suggesting the highest bendability, followed by TRIP780LD, DP980TD and DP980LD. The relationship between the spring-back and the bending angle, in these phenomena, the large of plastic deformation at higher bending angle on the bending area, this decreases the elongation property which in turn induces a lower spring-back angle and can be expressed in equation (1), where K is the ultimate tensile strength, θ is the bending angle, ρ is the position of the neutral axial, υ is the Poisson ratio, ω is the die gap, E is the Young's modulus and t is the thickness [6].

$$\frac{\Delta\theta}{\theta} = \frac{3K_{\rho}(1-\upsilon^2)_{(1+4t/\omega)}}{Et} \tag{1}$$

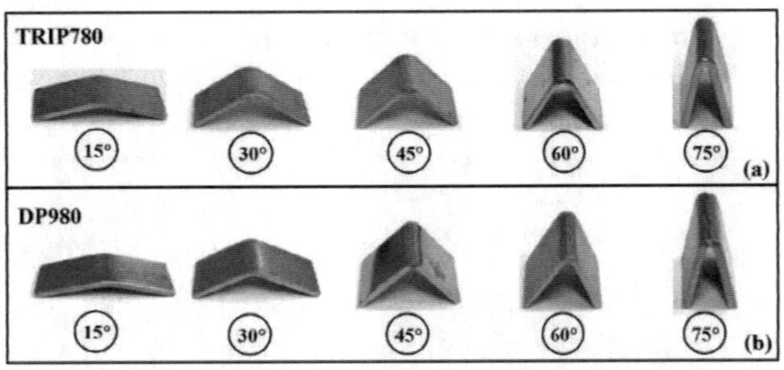

Figure 4. The LD-rolling failure mechanisms under various bending angles: (a) TRIP780 (b) DP980 steels.

Figure 4 compares the LD-rolling failure mechanisms under the bending angles of 15°-75° of the TRIP780 and DP980 steel sheets. In figure 4(a), the failure (i.e. cracks) started to appear on the roof of the TRIP780 sheet under the 75° bending angle condition and the crack propagation increased with increase in the bending angle but with no complete separation. Meanwhile, in figure 4(b), at the 45° bending angel, numerous crack lines emerged on the roof of DP980 and deteriorated as the bending angle increased. The total separation of the DP980 specimen occurred once the bending angle reached 75°.

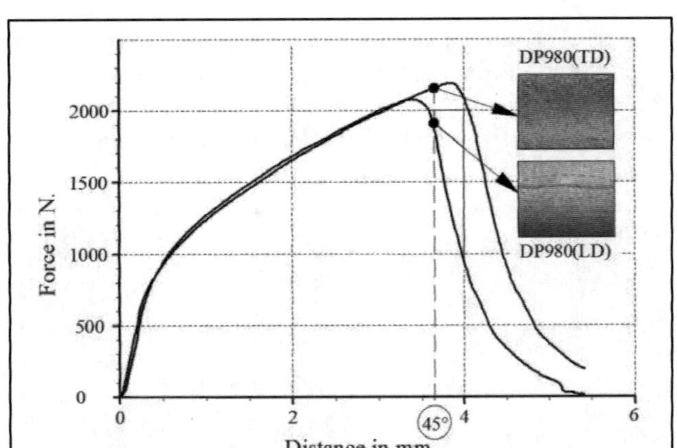

Figure 5. The relationship between the punch depth, rolling directions and bending force at the roof of DP980 sheet.

Figure 5 illustrates the relationship between the punch depth, rolling directions and bending force at the outer bent surface (roof) of DP980 sheet. In the figure, the bending force increased with the

IOP Conf. Series: Materials Science and Engineering **361** (2018) 012015 doi:10.1088/1757-899X/361/1/012015

punch depth (mm) for both rolling directions. The bending force belonging to the LD-rolling DP980 started to decline approaching (before) the 45° bending angle. On the other hand, in the traverse-rolling direction, the bending force declined shortly after arriving at the 45° bending angle. The forces continued to decrease as the punch depth increased. The findings indicated that the TD-rolling DP980 outperformed the LD-rolling counterpart with regard to the bendability and strength. The experimental findings are consistent with the theory of anisotropic materials [7] in which the traverse-rolling direction (TD) material exhibits greater bendability than that of the longitudinal-rolling direction (LD).

3.2. Failure Mechanisms

The failure mechanisms of the TRIP780 and DP980 steel sheets in the longitudinal direction (LD) of rolling were investigated and compared. As previously mentioned, the LD rolling is more susceptible to cracking and thereby the study focus. Figure 6 depicts the cross-sectional SEM failure mechanism of the LD-rolling TRIP780 steel under the 75° bending angle condition, where (A), (B) and (C) represent the outer, central and inner regions. Under the 75° bending angle, the failure was noticeably pronounced from the edge of the outer region to the central region, while small cracks appeared in the inner region. Upon closer inspection, a small crack however emerged and could be detected in the LD-rolling TRIP780 steel under the bending angle of 60°, as illustrated in figure 7.

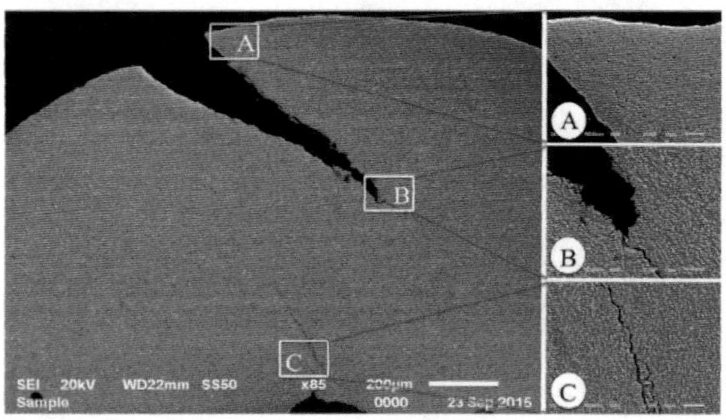

Figure 6. The cross-sectional SEM failure mechanism of the LD-rolling TRIP780 steel with the 75° bending angle: (A) the outer, (B) central, (C) inner regions.

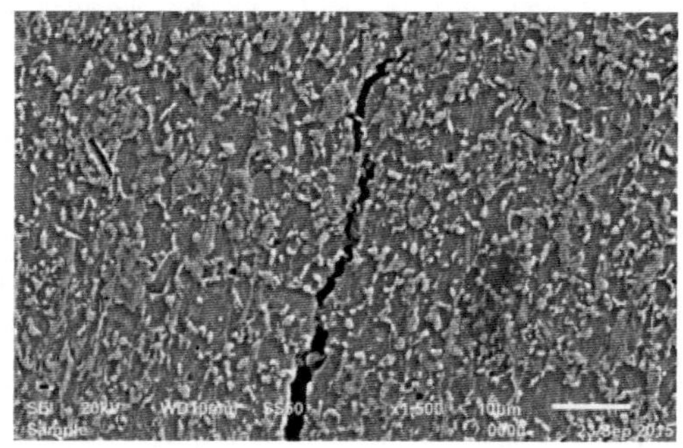

Figure 7. The cross-sectional OM of LD-rolling TRIP780 in which the initial crack appeared on the inner region with the 60° bending angle.

Figure 7 depicts the cross-sectional optical microscopy of the LD-rolling TRIP780 metal sheet under the 60° bending angle condition. In the figure, despite the outward compressive stress-induced hairline fracture on the inner region [8], the overall bendability of the metal sheet was minimally restricted. This phenomenon is attributable to a delayed crack propagation on this particular region relative to that of the tensile stress-induced outer region.

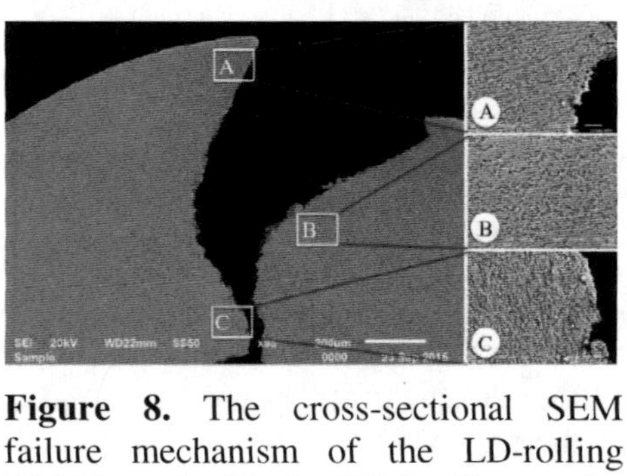

Figure 8. The cross-sectional SEM failure mechanism of the LD-rolling DP980 steel with the 75° bending angle: (A) the outer, (B) central, (C) inner regions.

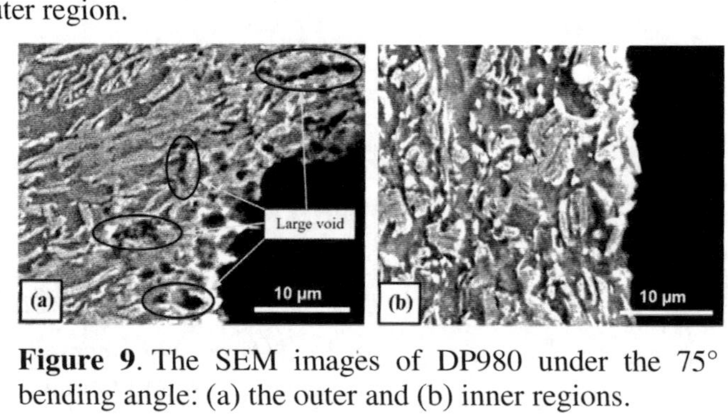

Figure 9. The SEM images of DP980 under the 75° bending angle: (a) the outer and (b) inner regions.

Figure 8 illustrates the cross-sectional SEM failure mechanism of on the outer (A), central (B) and inner (C) regions of the LD-rolling DP980 steel under the 75° bending angle, where a complete separation took place. In the figure, the crack propagation was of 45° in relation to the punch, consistent with Rèche et al [5]. Specifically, the phase in the outer region (A) was elongated along the outermost edge of the metal sheet, while there was minimal change in the martensitic phase in the central region (B). In the inner region (C), the shape of the martensitic phase altered as per the punch direction. In fact, the initial cracks on the DP980 steel sheet emerged under the bending angle 45° condition (Figure 4), with the extent of failure more pronounced on the outer than inner region. With the increase in the bending angle, the dual-directional cracks propagated to the central region of the metal sheet. A total separation of the specimen materialized at the 75° bending angle. Figure 9(a) and (b) are the magnified SEM images on the outer regions and the inner region, respectively, of the DP980 steel sheet under the 75° bending angle condition. In figure 9(a), voids of varying sizes (i.e. macro- and micro-voids) were present on the outer and central regions where ferrite and martensite interface, with the significant proportion residing on the outer region. The crack propagation on both the outer and central regions proceeded in the direction of the macro-voids inward (i.e. toward the inner region). Meanwhile, in figure 9(b), the crack propagation on this region (i.e. inner region) was attributable to the compressive pressure. Due to the compressive pressure, no void was detected on the inner region. The experimental results with regard to the initial crack of both metal sheet types are consistent with Pathak et al [9] and Zhao et al [10]. The results are in good agreement.

4. Conclusions

This experimental research has investigated the relationships between the bending angles and the bending force; and between the bending and spring-back angles of two types of advanced high strength steel (AHSS) sheets (TRIP780 and DP980) under two rolling direction conditions (traverse (TD) and longitudinal (LD)). The study also determined the effect of the bending angle on the failure mechanism of the metal sheets. The findings revealed that the bending force and the bending angle were positively correlated. Specifically, the bending force of TRIP780 steadily rose with increase in the bending angle under both rolling directions. Meanwhile, the bending force of the LD-rolling DP980 increased and decreased approaching the 45° bending angle while that of the TD-rolling DP980 increased and declined shortly after arriving at the 45° bending angle. Interestingly, the bending force of the LD-rolling DP980 reversed to zero at the 75° bending angle, the point at which a total separation occurred. On the other hand, the spring-back and bending angles were inversely correlated. The experiments revealed that the TD-rolling TRIP780 and LD-rolling DP980 sheets respectively had the largest and smallest spring-back angles under all the experimental bending angles. Nevertheless, the spring-back angle of the LD-rolling DP980 became zero at the 60° bending angle. On the failure mechanism, at the 45° bending angle, the initial crack could be detected on both the inner and outer regions of the LD-rolling DP980 sheet; and on the inner region of the TD-rolling TRIP780 sheet. The crack was more severe when the bending angle reached 60° for the TD-rolling TRIP780.

References

[1] Zhao P, Chang Y, Hu P and Wu Y 2016 *J. Mat. Pro. Tech.* 228 68–75.
[2] Rèche D, Sturel T, Bouaziz O, Col A and Gourgues-Lorenzon A F 2011 *Mater. Sci. Eng. A.* 528 5241–5250.
[3] Nakataki Y, Tsuyama S and Hosoya Y 2013 *Tetsu. to. Hagané.* **99** 245-253.
[4] Kaupper M and Merklein M 2013 *CIRP. Ann. Manuf. Techn.* 62 247–250.
[5] R'eche D, Besson J, Sturel T, Lemoine X and Gourgues-Lorenzon A F 2012 *Int. J. Mech. Sci.* 57 43–53.
[6] Fei D and Hodgson P 2006 *Nucl .Eng .Des.* 236 1847–1851.
[7] Panich S, Uthaisangsuk V, Suranuntchai S and Jirathearanat S 2014 *Mater. Sci. Eng. A.* 592 207-220.

[8] Aslam I, Li B, McClelland Z, Horstemeyer S J, Ma Q and Wang P T 2014 *Mater. Sci. Eng. A.* 590 168–173.
[9] Pathak N, Butcher C, Worswick M J, Bellhouse E and Gao J 2017 *Materials* 10, 346; doi:10.3390/ma10040346.
[10] Zhao Z, Huang S, Dan W, Zhang W and Li S 2017 *Mater. Sci. Eng. A.* 695 80–91.

Acknowledgements
The authors would like to express sincere gratitude to King Mongkut's Institute of Technology Ladkrabang (KMITL)'s Faculty of Engineering for the financial support.

IOP Conf. Series: Materials Science and Engineering **361** (2018) 012016 doi:10.1088/1757-899X/361/1/012016

Design of bevel gears using accelerated particle swarm optimization technique

Paridhi Rai, Asim Gopal Barman

Department of Mechanical Engineering, National Institute of Technology Patna, Patna 800005, Bihar, India

Email: asim@nitp.ac.in

Abstract. Straight bevel gears are simple type of bevel gears with their teeth being straight and tapered. They transmit power between shafts which are perpendicular with each other. In various applications of bevel gears, the issue of minimization of volume of gears is of immense importance for the designers, since it leads to reduced power consumption and material requirement for manufacturing. In this study, Accelerated Particle Swarm Optimization (APSO) is used to minimize the volume of straight bevel gears and optimized design variables are obtained. The design variables employed in the optimization procedure are module, face width and number of teeth. A design problem is considered for design optimization of bevel gear using APSO and obtained results are compared with the results produced by genetic algorithm (GA) and simulated annealing (SA) from previous literature. Obtained results show that there is improvement in the volume attained using APSO.

1. Introduction

Designing a gear drive is an intricate task which involves various factors such as gear geometry, heat treatment, manufacturing etc. in order to satisfy different practical considerations of high strength, high accuracy, low noise and small size. Several approaches are used by the researchers and designers in order to improve the gear design with respect to weight, volume, life, noise etc.

To obtain optimum gear design, numerous optimization techniques starting from traditional one to meta-heuristic have been used by the researchers in recent times. Golabi et al. [1] used non-linear programming technique to minimize the volume of one-, two- and three- stage gear train. Moving towards application of meta-heuristic optimization technique, Gologlu and Zeyveli [2] illustrated the use of Genetic Algorithm (GA) for finding minimum volume of two stage helical gearbox. Zhang et al. [3] proposed design optimization software based on genetic algorithm for obtaining optimized bevel gear design. Tudose et al. [4] used genetic optimization for performing optimization on two-stage co-axial helical gearbox. To obtain low weight and energy efficient helical gear pair, Tamboli et al. [5] used particle swarm optimization to minimize volume of a helical gear pair. In recent times, Zolfaghari et al. [6] designed a straight bevel gear using genetic algorithm (GA) and simulated annealing (SA). The results obtained from these two algorithms are then compared with traditional algorithm (TA). Panda et al. [7] introduced differential evolution algorithm for minimizing the weight of spur gear. Using obtained gear geometry parameters, a CAD model has been developed and stress analysis has been done using ANSYS. Straight bevel gears are utilized in transmission system where the axes intersect at 90°. Thus, they find applications where low velocity and production cost are desired [8]. Despite of this, availability of literature related to bevel gear volume optimization is

Content from this work may be used under the terms of the Creative Commons Attribution 3.0 licence. Any further distribution of this work must maintain attribution to the author(s) and the title of the work, journal citation and DOI.
Published under licence by IOP Publishing Ltd

limited. There are various PSO variants and hybrid algorithms, developed by combining other existing algorithms, which are increasingly popular and able to solve many design problems. But APSO is much simpler as it utilizes only two parameters and has faster convergence [9]. In this paper, volume of bevel gear is optimized using APSO and optimum design values are obtained. All the formulas and instructions are as per AGMA standards. A design example has been taken from literature [6] for optimization. The obtained results are validated by comparing it with previously published results.

Rest of the paper is organized in following manner. The next section consists of explanation of optimization algorithm i.e. APSO. In section three, mathematical modelling is done. A design example has been presented in section four. Simulation results of above problem formulated are reported in Section 5. Finally, conclusion is drawn in Section 6.

2.Accelerated Particle Swarm Optimization

Global best and individual best position of particles both are used in standard particle swarm optimization problem [10]. When individual best are used, there found to be diversity in the quality solution; however this diversity can be simulated using randomness. Consequently, there is no compelling condition to use individual best, until the objective function is very complex and non-linear in nature. By using global best, the algorithm can be accelerated to converge. A simple formula is employed to generate velocity vector, with $\boldsymbol{x}_i^t$ and $\boldsymbol{v}_i^t$ as the current position and velocity vector respectively, and is given by

$$v_i^{t+1} = v_i^t + \alpha\epsilon_n + \beta\left(G^* - x_i^t\right) \tag{1}$$

where ϵ_1 and ϵ_2 are the random numbers between 0 and 1. Here, α and β are the acceleration constant. The expression to update the position is same as used in standard PSO. To enhance the convergence even more, the update of position can be written as

$$x_i^{t+1} = (1 - \beta)x_i^t + \beta G^* + \alpha\epsilon_n \tag{2}$$

Thus APSO based optimization technique is quiet simple and results in quick convergence [9].

3.Mathematical Modelling

In this study, minimizing the volume of bevel gears is treated as objective function for the design optimization problem. For formulating the mathematical model, objective function, design variables and design constraints are required. In the following section, brief discussion has been made on problem formulation.

3.1 Design Variables

Design variables considered for optimization procedure are module, pinion teeth number and face width of gear. Description of these variables with their minimum and maximum bounds is given in Table 1.

Table 1. Description of Design variables

Design Variables	Min. Value	Max. value
Module (m)	1	8
Pinion teeth number (z_1)	14	60
Face width (b)	10	60

3.2 Formulation of Objective function

Volume is regarded as one the significant criteria for a gear reducer. Therefore, minimization of volume of bevel gear is considered as the objective function and expressed as.

$$V = \frac{\pi}{3} b\cos\delta_1 \left[\left(\frac{mz_1}{2}\right)^2 + \left(\frac{mz_1}{2}\right)\left(\left(\frac{r-b}{2}\right)\left(\frac{mz_1}{2}\right)\right) + \left(\left(\frac{mz_1}{2}\right)\left(\frac{r-b}{2}\right)\right)^2\right] +$$

$$+\frac{\pi}{3} b cos\delta_2 \left[\left(\frac{mz_2}{2}\right)^2 + \left(\frac{mz_2}{2}\right)\left(\left(\frac{r-b}{2}\right)\left(\frac{mz_2}{2}\right)\right) + \left(\left(\frac{mz_2}{2}\right)\left(\frac{r-b}{2}\right)\right)^2\right] \tag{3}$$

where δ_1 and δ_2 the cone angle for pinion and gear respectively and r is called cone pitch.

3.3 Constraints

Various design constraints are required for satisfying the practical demands such as bending strength, pitting resistance and linear velocity of gear at outer diameter of the design problem respectively. These established design constraints are expressed as shown below

$$g_1(x) = Z_E W_t K_v K_{H\beta} K_A K_\theta^2 Z_z^2 S_H^2 Z_{XC} - (\sigma_{allowH} Z_{NT} Z_w)^2 b d Z_1 \leq 0 \tag{4}$$

$$g_2(x) = K_v W_t K_{H\beta} K_A K_\theta Y_X S_F - \sigma_{allowF} Y_{NT} Z_J b m_{et} Y_\beta \leq 0 \tag{5}$$

$$g_3(x) = 1047.2(10^{-6}) m Z_1 n_l - \left(50 + 56\left(1 - 0.25(12 - Q_V)^{2/3}\right) + (Q_V - 3)\right)^2 \leq 0 \tag{6}$$

$$g_4(x) = b \leq \{0.3A_m, 10m_{et}\} \tag{7}$$

where $K_v, K_{H\beta}, K_\theta$ *and* K_A is the dynamic, load distribution, temperature and overload factor respectively. W_t is the transmitted load. Here S_F and S_H is safety factor for bending and contact and m_{et} is the outer transverse module. Stress cycle factor for bending and pitting is denoted by Y_{NT} and Z_{NT} respectively and Y_X and Z_X is the size factor bending and pitting resistance respectively. Y_β, Z_w, Z_{XC} *and* Z_z length-wise curvature, hardness, crown and reliability factor respectively for bevel gears. Z_J *and* Z_E are bending geometry factor and elastic coefficient for pitting respectively. All the constants used in above inequalities are explained and illustrated in [6] and [8]. To prevent undercut in straight bevel gears, minimum number of teeth required are 14. Furthermore, based on literature [6], following equalities are considered in order to avoid undercutting.

$$g_{eq1}(x) = A_{iG} = A_m - 0.5b \tag{8}$$

$$g_{eq2}(x) = \psi_{iG} = 0 \tag{9}$$

$$g_{eq3}(x) = \delta_p = \delta_G \tag{10}$$

$$g_{eq4}(x) = \tan\varphi_{Ti} = \tan\varphi \tag{11}$$

$$g_{eq5}(x) = b_{ilp} = A_{iG} \tan\gamma \sin^2\varphi \tag{12}$$

$$g_{eq6}(x) = b_{ip} = b_p \tag{13}$$

Here A_{iG} *and* A_m are gear cone distance and mean cone distance (mm) respectively. ψ_{iG} represents the inner gear spiral angle. φ_{Ti} *and* φ denote the transverse and normal pressure angle respectively. b_{ilp}, b_{ip} *and* b_p is the pinion limit inner dedendum, pinion mean inner dedendum and pinion inner dedendum respectively. δ_p *and* δ_G is pinion and gear dedendum angle respectively.

4. Design Problem

In this study, the optimal design values of bevel gear have been achieved using APSO in MATLAB. For designing bevel gear pair, various design parameters and working condition required for optimization of bevel gears are shown in Table 2.

Table 2. Various design parameters used in design optimization problem [6]

Design Parameters	Values
Power transmitted (kW)	5
Reduction ratio	1:1
Input speed (rpm),	900
Load on prime mover	Medium shock
Contact safety factor	1.2
Bending safety factor, σ_{allowF}	1.2
Pressure angle, φ	20°
Material	Steel (Grade II)
Hardness (HB) (Gear/Pinion)	350
Transmission Accuracy number, Q_V	11
Life (No. of cycles)	10^8

Load of driven machine	Medium shock
Reliability	0.99
Installation condition	Both members straddle mounted
Gear/pinion shape	Properly crowned teeth
Tooth form	Full depth

5. Results and Discussion

To produce optimum results, APSO algorithm has been coded in MATLAB. Maximum number of iterations is taken as stopping criteria for optimization. The typical values of acceleration constants are assumed to lie between: $\alpha \cong 0.1$ to 0.4 and $\beta \cong 0.1$ to 0.7. Description of various parameter used in APSO are reported in Table 3.

Table 3. Various APSO parameters used for optimization of bevel gears

APSO parameters	Values
Population size	20
Iterations	150
Time steps	250
α	0.2
β	0.5

A comparison of results has been shown in Table. 4 The effectiveness of transmitted power on the volume of straight bevel gear is highlighted in the table. Convergence curve for volume vs iterations is shown in Figure 1. As demonstrated in Table 4, APSO led to improvement in gear volume from 1.12% to 29.9% for all transmitted power (Figure 2). The face width values are increasing from 5 kW power to 30.65 kW power, but the obtained values are smaller in comparison with the earlier values obtained from other algorithms. This will help in quiet, smooth and reliable gear mesh action. Module of bevel gear increases with increase in transmitted power value. Finally, the numbers of teeth increases and then decreases as power transmitted varies from 5 kW to 30.65 kW (Figure 3).

Table 4. Comparison of results for APSO with TA, GA and SA [6]

Power(kW)	5				10.65				30.65			
Design Variables	TA	GA	SA	APSO	TA	GA	SA	APSO	TA	GA	SA	APSO
Module, m	3.5	2.75		4.0	4.5	5.5	4.5	4.5	5.5	5.5	5.5	5.5
No. of teeth, (z_1/z_2)	25	20	25	25	30	15	20	27	35	20	30	20
Face width, (mm)	19.2	27.3	18.5	15.86	23.1	26.6	22.8	17.76	26.9	46.02	38.2	22.47
Volume, dm^3	69.8	63.3	68.7	51.06	206.8	158.9	170.8	144.9	502	453.6	453.8	448.59
% Improve. by APSO	26.8	19.3	25.6	–	29.9	8.8	15.1	–	10.6	1.12	1.16	–

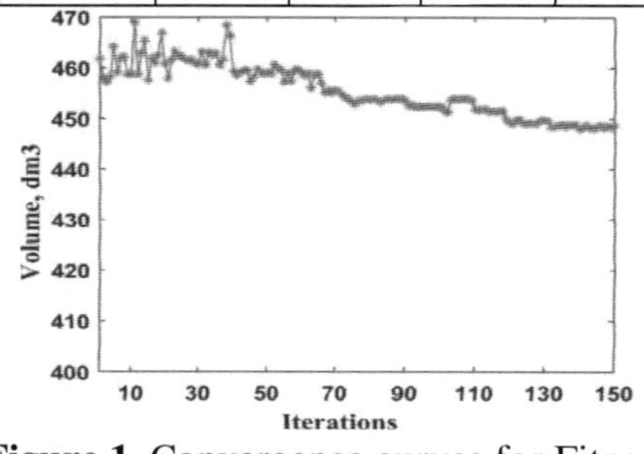

Figure 1. Convergence curves for Fitness Function vs iterations

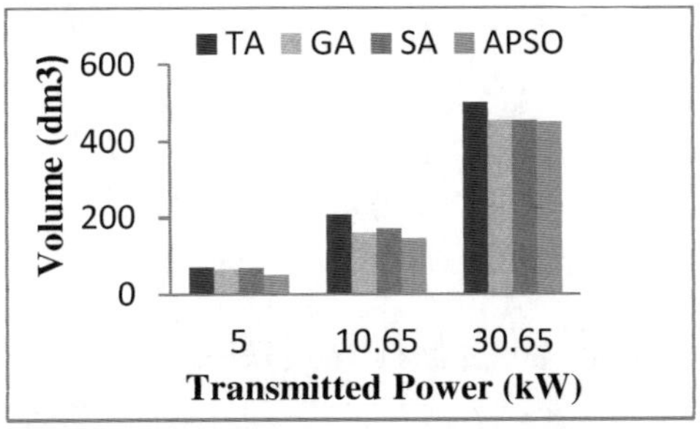

Figure 2. Optimized volume of bevel gear compared for TA, GA, SA and APSO

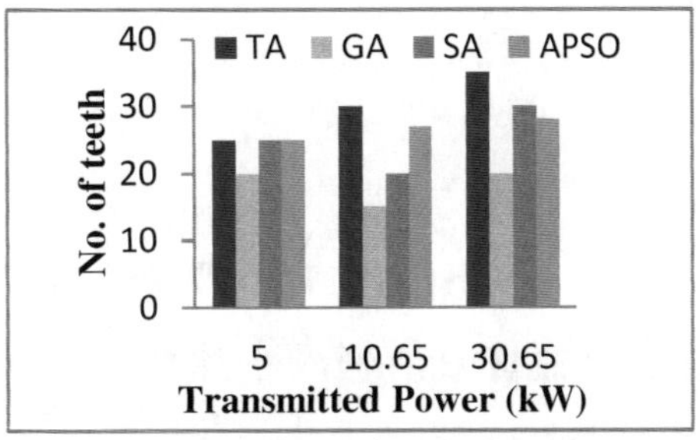

Figure 3. Variation of No. of teeth vs transmitted power for TA, GA, SA and APSO

6. Conclusion

In this study, accelerated particle swarm optimization is used for volume optimization of bevel gears. The optimal gear volume achieved using APSO is found to be better than earlier used optimization techniques (TA, GA and SA). It has been found that the values of face width are smaller in comparison to previously published work. This will make gears less sensitive to errors of manufacturing and misalignment. There is slight change in value of gear module using APSO. Number of teeth obtained from APSO is greater than that obtained from GA, but less for SA. As the value of transmitted power changes, the percentage of reduction in volume of bevel gear when compared to other algorithms lie in between 1.12% to 29.9%. Hence the achieved results show that proposed APSO approach surpasses the other approaches in minimizing volume of bevel gears. This will reduce material requirement for manufacturing which will further reduce the manufacturing cost.

References

[1] S. Golabi, J.J. Fesharaki, and M. Yazdipoor, 2014, Gear train optimization based on minimum volume/weight design, *Mechanism and Machine Theory*, **73**, pp 197-217.

[2] C. Gologlu and M. Zeyveli, 2009, A genetic approach to automate preliminary design of gear drives, *Computers & Industrial Engineering*, **57,** pp. 1043-1051.

[3] X Zhang, Y Rong, J Yu, L Zhang and L Cui, 2011, Development of optimization design software for bevel gear based on integer serial number encoding genetic algorithm, *JSW*, **6**, no 5.

[4] L. Tudose, O. Buiga, C. Stefanache, 2010, Automated optimal design of a two-stage helical gear reducer, *Structural Multidisciplinary Optimization*, **42**, No. 3, pp 429–435.

[5] K. Tamboli, S. Patel, P. M. George and R. Sanghvi, 2014, Optimal Design of a Heavy Duty Helical Gear Pair Using Particle Swarm Optimization Technique, *Procedia Technology*, **14**, pp 513-519.

[6] A. Zolfaghari, M. Goharimanesh and A.A. Akbari, 2017, Optimum design of straight bevel gears pair using evolutionary algorithms, *Journal of the Brazilian Society of Mechanical Sciences and Engineering*, **39**, No. 6, pp. 2121-2129.

[7] S. Panda, B.B. Biswal, S.D. Jena and D. Mishra, 2017, An approach to weight optimization of a spur gear, *Proc. of the Inst. of Mech Eng Part J: Journal of Eng Tribology*, **231**, pp 189-202.

[8] A. Standard, 2005, Design manual for bevel gears, *ANSI/AGMA*, pp. 2005-D03.

[9] P. Kunche, K.V. Reddy, 2016, Speech Enhancement Approach Based on Accelerated Particle Swarm Optimization (APSO), *Metaheuristic Applications to Speech Enhancement* pp. 39-60.

[10] X.S. Yang, S. Deb and S. Fong, 2011, Accelerated particle swarm optimization and support vector machine for business optimization and applications, *Networked digital technologies*, pp.53-66.

IOP Conf. Series: Materials Science and Engineering **361** (2018) 012017 doi:10.1088/1757-899X/361/1/012017

Analysis of Cooling Performance of a Marine Air-cooled Sliding Intermediate Bearing Based on Fluid-Structure Interaction

Shao Yong

China Ship Development and Design Center, Wuhan 430064, China

Email: shaoyong1111@163.com

Abstract. The intermediate bearing is an important component in marine propulsive system. The transmission efficiency, reliability and service life of propulsion shafting are significantly affected by the lubrication performance of intermediate bearing. Based on the theory of CFD, the FSI heat transfer numerical model of a marine air-cooled sliding intermediate bearing was established. The hydrodynamic lubrication numerical model of intermediate bearing was coupled with the heat transfer model by UDF. The temperature fields of intermediate bearing and oil in tank were obtained under the maximum speed condition. The simulation results agree well with the experiment and the accuracy of numerical model is also accepted at the same time. As the results shown, the highest temperature of bearing bush is 77.88°C and the average temperature of oil in tank is 70.14°C. The results indicate that the cooling performance of intermediate bearing meets the design requirements.

1. Introduction

The propulsive shafting is the important equipment in marine power system and its stability and reliability have a direct impact on ship's dynamic performance. The intermediate bearing is mainly used to support the rotation of propulsive shafting, which endures heavy loads in the working conditions. The lubrication performance of intermediate bearing is affected by many factors. Among them, the temperature of oil is one of the primary factors [1]. Therefore, the study on the cooling performance of intermediate bearing plays a significantly important role in the improvement of its service performances.

In this paper, the hydrodynamic lubrication model and the FSI (Fluid-Structure Interaction, FSI) heat transfer model of intermediate bearing were established respectively [2-3]. In order to take account of the interactive effects, the above two numerical models were coupled by UDF (User Define Function, UDF) supported by the commercial CFD (Computational Fluid Dynamics, CFD) software FLUENT. The study on the cooling performance was carried out under the maximum speed condition.

2. Basic Theory

2.1. Lubrication Theory

2.1.1. Lubrication Governing Equation. The basic governing equation of hydrodynamic lubrication model is the Reynolds equation. Based on the different assumptions, the Reynolds equation can be

Content from this work may be used under the terms of the Creative Commons Attribution 3.0 licence. Any further distribution of this work must maintain attribution to the author(s) and the title of the work, journal citation and DOI.
Published under licence by IOP Publishing Ltd

obtained in different forms. In this paper, the hydrodynamic lubrication models are established on the average Reynolds equation with the consideration of surface roughness [4-5]. The basic lubrication governing equation is shown in equation (1).

$$\frac{\partial}{\partial x}(\phi_x \frac{\rho h^3}{\mu}\frac{\partial p}{\partial x})+\frac{\partial}{\partial y}(\phi_y \frac{\rho h^3}{\mu}\frac{\partial p}{\partial y})=6U\rho\phi_c\frac{\partial h}{\partial x}+6U\sigma\frac{\partial(\rho\phi_s)}{\partial x} \quad (1)$$

Where, h is the nominal oil film thickness. p is the oil film pressure. x is the circumferential direction of the bearing. y is the axial direction of the bearing. U is the relative velocity of journal and bearing bush in the circumferential direction. μ is the viscosity of oil. ρ is the density of oil. ϕ_x and ϕ_y are the pressure flow factors in x and y direction on the lubrication surface respectively. ϕ_s is the stress flow factor.

2.1.2. Oil film thickness equation. Figure 1 shows the oil film thickness h of a sliding bearing [6]. O_1 and O_2 are the center of the bearing and journal separately. R_1 and R_2 are the radius of bearing and journal. ϕ is the offset angel. θ is the circumferential coordinates of the film thickness, which take the point at the maximum oil film thickness as the origin.

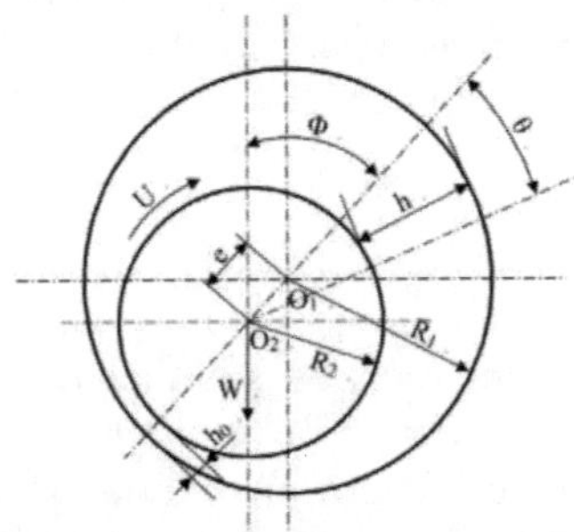

Figure. 1 Oil film thickness of a sliding bearing

According to the above geometrical relations, the oil film thickness h can be deduced as equation (2), shown as below.

$$h=c+e\cos\theta \quad (2)$$

Where, c is the radius clearance defined by $c=R_1-R_2$. e is the eccentricity defined by $e=R_1-R_2-h_0$. h_0 is defined as the minimum oil film thickness.

2.2. CFD Basic Theory

The law of mass, momentum and energy conservation must be obeyed in fluid dynamics [7]. To describe different flowing phenomena, several essential flow equations should be added.

2.2.1. The continuity equation. The continuity or mass-conservation equation must be obeyed in any flowing situation as Equation (3) shows.

$$\frac{\partial\rho}{\partial t}+\frac{\partial(\rho u_x)}{\partial x}+\frac{\partial(\rho u_y)}{\partial y}+\frac{\partial(\rho u_z)}{\partial z}=0 \quad (3)$$

Where, u_x, u_y and u_z are the velocity component in the direction of x, y and z respectively. t is the time.

2.2.2. The momentum equation. The momentum equation is based on Newton's second law, which means as a differential unit, the changing rate of momentum with respect to time is equal to the total forces acting on its surfaces.

$$\frac{\partial(\rho u_x)}{\partial t}+\nabla\cdot(\rho u_x\vec{u})=-\frac{\partial p}{\partial x}+\frac{\partial\tau_{xx}}{\partial x}+\frac{\partial\tau_{yx}}{\partial y}+\frac{\partial\tau_{zx}}{\partial z}+\rho f_x \quad (4)$$

$$\frac{\partial(\rho u_y)}{\partial t}+\nabla\cdot(\rho u_y\vec{u})=-\frac{\partial p}{\partial y}+\frac{\partial \tau_{xy}}{\partial x}+\frac{\partial \tau_{yy}}{\partial y}+\frac{\partial \tau_{zy}}{\partial z}+\rho f_y \quad (5)$$

$$\frac{\partial(\rho u_z)}{\partial t}+\nabla\cdot(\rho u_z\vec{u})=-\frac{\partial p}{\partial z}+\frac{\partial \tau_{xz}}{\partial x}+\frac{\partial \tau_{yz}}{\partial y}+\frac{\partial \tau_{zz}}{\partial z}+\rho f_z \quad (6)$$

Where, p is the pressure on the unit. τ_{xx}, τ_{xy} and τ_{xz} are the component of viscous stress. τ resulting from the viscosity effect among molecules. f_x, f_y and f_z are the unit mass force in three directions.

2.2.3. The energy equation. The law of energy conservation which based on Newton's first law must be obeyed by the flowing systems including heat transfer. According to the law, the rate of energy increasing per unit is equal to the sum of net heat flux added into the unit and the work that the mass force and surface force has done. The energy equation is shown as below.

$$\frac{\partial \rho E}{\partial t}+\nabla\cdot\left[\vec{u}\left(\rho E+p\right)\right]=\nabla\cdot\left[k_{eff}\nabla\cdot T-\sum h_j J_j+\tau_{eff}\cdot\vec{u}\right]+S_h \quad (7)$$

Where, E is the total energy including the internal energy, kinetic energy and potential energy. h_j is the enthalpy of the component j. k_{eff} is the effective heat conductivity. J_j is the diffusion flux of component j. S_h is the source term including the reaction heat and other heat defined by users.

3. Numerical Modeling

3.1. Geometry Model

The air-cooled sliding intermediate bearing composes of different parts. The main components include bearing cap, bearing block, upper/lower bearing bush, upper/lower oil tank, oil flinger, oil collector, connecting bolts and so on. The three dimensional models of above parts are established and assembled according to the containment relationships. The assembly model of intermediate bearing is shown as Figure 2.

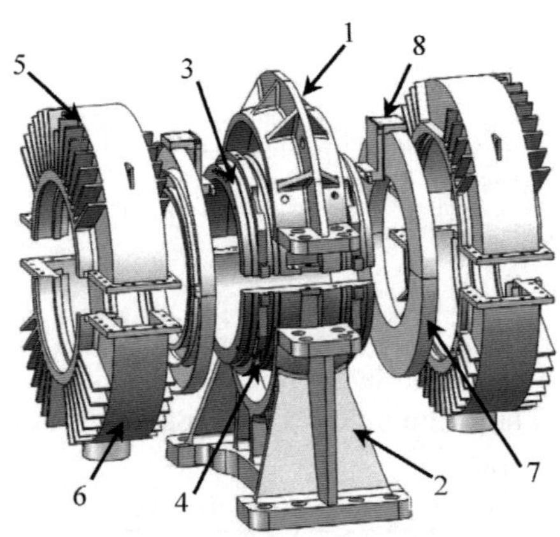

1-bearing cap; 2-bearing block; 3-upper bearing bush; 4-lower bearing bush; 5-upper shell of oil tank; 6-lower shell of oil tank; 7-oil flinger; 8-oil collector

Figure. 2 Geometrical models of intermediate bearing

3.2. Mesh Generation

Two oil tanks are set on the left and right side of the intermediate bearing, in which an amount of oil have been stored. The oil gathers on the bottom of oil tank and forms the oil pool with a certain depth. The oil flinger is equipped in each oil tank and its bottom immerses in the oil pool. According to the geometrical configuration, the geometry model of oil pool in tank can be obtained.

IOP Conf. Series: Materials Science and Engineering **361** (2018) 012017 doi:10.1088/1757-899X/361/1/012017

Based on the CFD software FLUENT, the solid structures of intermediate bearing and the fluid zone of oil pool in tank have been meshed respectively with the unstructured tetrahedral mesh.

3.3. Boundary Conditions

The oil flinger of intermediate bearing is fixed on the propulsive shaft and rotates along with the shaft. A certain amount of oil in tank will be thrown up when the oil flinger rotates. Most of them is collected by the oil collector and then transported to the bearing bush. The oil in bearing bush not only can lubricate bush and reduce the friction force, but also can take away most of the friction heat generated at the bearing bush, which finally is taken to the oil pool in tank. The rest of oil thrown up by the oil flinger directly returns to the oil pool. The heat convection in the pool and the heat transfer between oil pool in tank and the lower shell of tank are effectively enhanced by the stirring of oil flinger. The heat imported from the internal walls of oil tank and bearing bush transfers among the parts of intermediate bearing in the way of heat conduction. The heat dissipation is achieved by the heat convection and the heat radiation on the exterior surfaces of intermediate bearing. The thermal equilibrium of intermediate bearing is reached by the comprehensive effects of above heat transfer methods in working condition.

Considering the features of heat transfer listed above, the following simplifications are introduced in FSI heat transfer model of intermediate bearing.

(1) The heat convection and radiation exist on the exterior surfaces of intermediate bearing. Since the maximum temperature of the exterior surfaces is below 65°C, the heat radiation is neglected. The external flow model of intermediate bearing is established and the convection heat transfer coefficients of each exterior surfaces are obtained. The third boundary conditions composed of the external temperature T_0 and the convection heat transfer coefficients α are applied on the exterior surfaces of intermediate bearing and the heat convection between the intermediate bearing and environment is correspondingly considered.

(2) Since the intermediate bearing is installed on the bearing pedestal, the bottom of the bearing block is defined as the constant environmental temperature T_0.

(3) The heat conduction exists among all the parts of the intermediate bearing. Hence, the contact surfaces of the parts are defined as the thermal-coupling wall.

(4) The heat flux q is applied on the inner surface of bearing bush accordingly [8]. In this way, the friction heat is considered.

(5) The process of oil thrown up from the oil pool by oil flinger is ignored. In order to consider the effects caused by oil flinger on the heat convection of oil pool, the contact surfaces between the oil pool and oil flinger are defined as a wall with the rotational velocity in working condition.

(6) The interfaces between the oil pool and the tank are defined as the thermal-coupling walls, which realize the coupled heat transfer between solid and fluid zones.

(7) As Figure 3 shown, one outlet and two inlets of oil are set on the upper surface of oil pool. The outlet, defined as a pressure outlet, indicates the oil thrown up from the pool by oil flinger. The inlets are defined as mass-flow inlets. The inlet 1 indicates the oil which is collected and transported to the bearing bush and finally flows back to the oil pool. The inlet 2 indicates the rest of oil which is thrown up by the oil flinger and directly falls down to the oil pool.

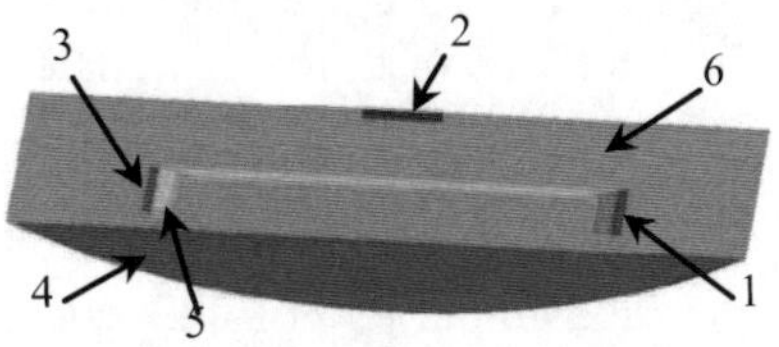

1-outlet; 2-inlet 1; 3-inlet 2; 4-interface between oil pool and oil tank; 5-contact surfaces between oil pool and oil flinger; 6-upper surface of oil pool

Figure. 3 Boundary conditions of the oil pool model

According to simplifies and assumptions, the FSI heat transfer model of the air-cooled sliding immediate bearing is established. Table 1 shows the main boundary conditions in this numerical model.

Table. 1 Main B.C. of FSI heat transfer model

Zone	Boundary type	Specifications
Exterior surfaces	wall	HTC α, Environment Temp. T_0
Bottom surfaces	wall	Constant Environment Temp. T_0
Loading surface of friction heat	wall	Heat flux, q
Contact surfaces among solid constructs	coupled-wall	Coupled heat transfer
Oil inlet 1 of oil pool	mass-flow inlet	Mass flow rate, Q_{m1}; T_1
Oil inlet 2 of oil pool	mass-flow inlet	Mass flow rate, Q_{m2}, T_2
Oil outlet of oil pool	pressure outlet	Environment pressure, P_0
Contact surfaces between oil pool and oil flinger	wall	Rotational angular velocity, ω
Interfaces between oil pool and oil tank	coupled-wall	Coupled heat transfer

3.4. Coupled Model Based on UDF

The thermal load applying to the intermediate bearing, which result from the friction heat generated in bearing bush, is affected by variable factors, such as rotational speed, exterior load and oil properties. Especially, the friction heat is significantly affected by oil temperature. The interactions between the oil temperature and thermal load applying to the intermediate bearing must be taken into account in numerical model.

Based on UDF function applied by commercial software FLUENT, the FSI heat transfer model of the sliding intermediate bearing is coupled with the hydrodynamic lubrication model. Figure 4 shows the framework of the coupling model.

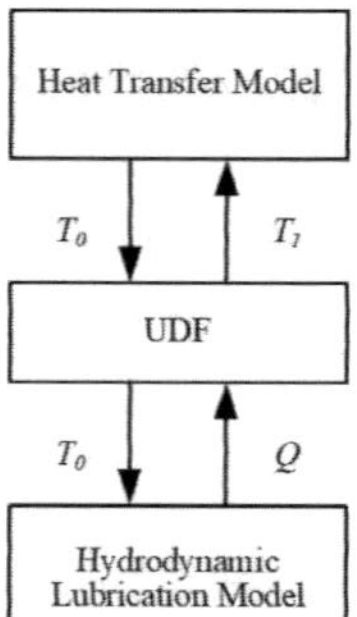

Figure. 4 Framework of coupling model based on UDF

As Figure 3.3 shown, data transfer between two models are realized by UDF, which acts as the bridge between FSI heat transfer model and lubrication model. In the process of calculation, the oil temperature of the oil pool outlet is firstly read from the FSI heat transfer model and then transferred to the lubrication model by UDF, which is defined as the temperature of oil supplied to the bush. According this oil input temperature for lubrication model, the lubrication performance is analyzed and the friction heat is calculated. Based on the friction heat, the temperature of oil inlet 1 defined as T_1 and the heat flux loaded to the inner surface of bush are updated. Then, the next cycle of iterative calculation begins until the results are convergent.

3.5. Material Parameters

Several materials are involved into the simulation of the cooling performance, including solid materials of each part in intermediate bearing and fluid material of the oil. The parameters of solid materials are obtained from the material manual. The physical parameters of the oil are obtained from test data. The details are specified in Table 2 and Table 3.

IOP Conf. Series: Materials Science and Engineering **361** (2018) 012017 doi:10.1088/1757-899X/361/1/012017

Table. 2 Parameters of solid materials

Type	Density kg/m3	Thermal Conductivity Coefficient W/(m.K)	Specific Heat J/(kg.K)
ZG230-450H	7830	50.7	469
20#(Forged Steel)	7850	48	469
ZL101A(T6)	2680	150	380

Table. 3 Properties of oil (20℃)

Property	Value	Unit
Density	891.6	kg/m^3
Specific heat	1953.1	J/(kg·K)
Thermal conductivity coefficient	0.144	W/(m·K)
Kinematic viscosity	484.0	mm^2/s

4. Results and Discussion

4.1. Lubrication Performance

The lubrication performance of the intermediate bearing under the maximum speed condition was investigated in the case that the oil temperature is 65°C. The calculation results are listed in Table 4.

Table. 4 Results of lubrication performance

Item	Unit	Value
Min. oil film thickness	μm	57.20
Min. oil film thickness ratio	—	63.95
Offset angle	°	33.34
Friction force	N	443.95
Friction coefficient	—	0.00222
Total friction power loss	kW	2.3478

4.2. Cooling Performance

The cooling performance of intermediate bearing under the maximum speed condition was investigated. The results are listed as follow. Figure 5 shows the temperature fields of main solid parts and cross-section of temperature and velocity of oil pool in tank.

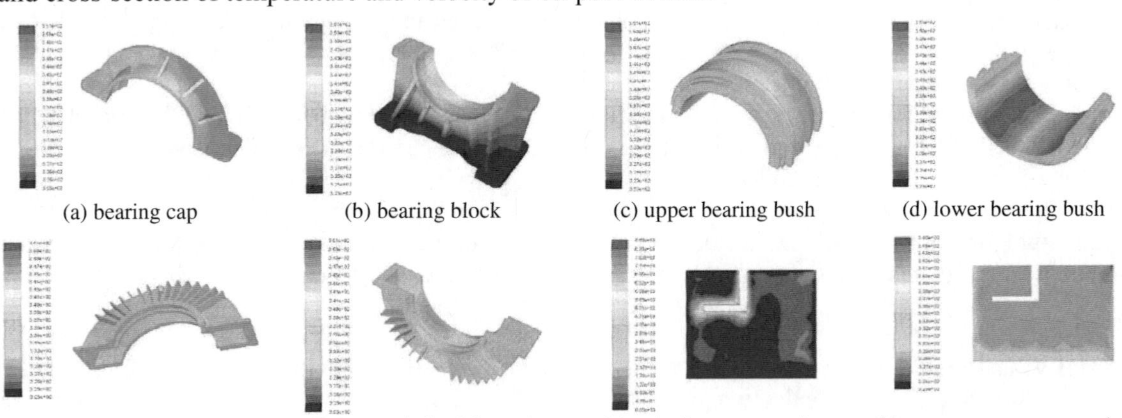

(a) bearing cap (b) bearing block (c) upper bearing bush (d) lower bearing bush

(e) upper shell of the tank (f) lower shell of the tank (g) velocity cross-section (h) temperature cross-section

Figure. 5 Contours of cooling performance

As the above results shown, the maximum temperature of the lower bearing bush is 77.88°C and the average temperature of oil pool in tank is 70.14°C under the maximum speed condition, which satisfied the technical requirements.

5. Comparison with Experimental Results

The cooling performance test platform of intermediate bearing has been set up as figure 6. The exterior load applied on the intermediate bearing is controlled by the hydraulic system and the adjustable-speed motor is used to control the rotational speed of the journal. During the test, the adjustment of the working condition is utilized by the central console.

Figure. 6 Test platform of intermediate bearing

The test of cooling performance was accomplished under the maximum speed condition. The results of test points in the intermediate bearing obtained from the simulation and test are listed in Table 5. From the comparison, it can be seen that the results from simulation agree well with the test data. The maximum error is below 5%, which indicates the accuracy of the model is acceptable.

Table. 5 Results of simulation and experiment

Test Point	Exp. Value /℃	Sim. Value /℃	ΔT /℃	Error /%
Bearing bush	70.5	70.31	0.19	0.27
Oil pool in tank	69.9	66.61	3.29	4.71

6. Conclusions

In this paper, the cooling performance of an air-cooled sliding intermediate bearing was analyzed. The conclusions can be drawn as follows:

(1) Based on the relative theory, the FSI heat transfer model and lubrication model are established separately. Two models are coupled based on UD. Hence, the lubrication performance is taken account into the heat transfer model.

(2) The cooling performance test platform of intermediate bearing has been set up. By contrast, the results from simulation agree well with test data. The maximum error is below 5%, which indicates the accuracy of the model is acceptable.

(3) As the results shown, the maximum temperature of bearing bush is 77.88°C and the average temperature of oil pool in tank is 70.14°C under the maximum speed condition, which meet the technical requirements. The cooling performance of intermediate bearing is satified.

References

[1] F Wlassow, F Duchaine, G Leroy and N Gourdain 2010 3D Simulation of Coupled Fluid Flow and Solid Heat Conduction for the Calculation of Blade Wall Temperature in a Turbine Stage *ASME Turbo Expo (Power for Land, Sea, and Air)* pp 723–734

[2] Q Liu, E A Luke and P Cinnella 2015 Coupling Heat Transfer and Fluid Flow Solvers for Multidisciplinary Simulations *Journal of Thermophysics & Heat Transfer* vol 19(4) pp 417–427

[3] G Chen and J B Heywood 1987 Numerical Analysis of Unsteady Heat Transfer in Coupling System of Combustion Chamber *Journal of Huazhong University of Science & Technology* vol 4 pp 29–36

[4] X Ye 2004 Numerical Investigation and Application of Three-Dimensional Lubrication

Performance in Piston Ring Pack *Huazhong University of Science & Technology*

[5] R Santhosh, J L, Hee, Simmons, K G Johnson, D Hann and M Walsh 2017 Experimental Investigation of Oil Shedding From an Aero-Engine Ball Bearing at Moderate Speeds *ASME Turbo Expo (Turbomachinery Technical Conference and Exposition)* pp V07AT34A018

[6] Q Lin, Z Wei, W Ning and W Chen 2013 Analysis on the lubrication performances of journal bearing system using computational fluid dynamics and fluid–structure interaction considering thermal influence and cavitation *Tribology International* vol 64 pp 8–15

[7] N P Starostin, A S Kondakov and M A Vasilieva 2013 Identification of friction heat generation in sliding bearing by temperature data *Inverse Problems in Science and Engineering* vol 21.2 pp 298–313

[8] N Tala-Ighil and M Fillon 2015 A numerical investigation of both thermal and texturing surface effects on the journal bearings static characteristics *Tribology International* vol 90 pp 228–239

IOP Conf. Series: Materials Science and Engineering **361** (2018) 012018 doi:10.1088/1757-899X/361/1/012018

Effect of welding sequence on welding deformation in seal cylinder

W Zhang, T Jiang, L Q Liu

Changchun University of Science and Technology, Changchun 130022, China

Email: zw0704@163.com

Abstract. The method of numerical simulation and experiment is used to calculate and analyze the post-weld deformation. The results show that the numerical simulation can quickly calculate the amount of welding deformation under different welding sequence, different welding meat size and different heat input, so as to optimize the welding process, control and reduce the welding deformation. The finite element results are less than the experimental results, which verified the accuracy of finite element calculation.

1. Introduction

When the manufacturing error appear in the work-piece surface, the surfacing[1,2,3] is usually selected to build welding in order to meet the size requirements, and improve the strength of the work-piece as the surface of the work-piece meet the corrosive media. In many projects, the welding area is often a complex surface, so surfacing quality has very strict requirements. If the welding quality is good, you can omit the machining process, otherwise it will make the manufacturing process more complicated, resulting in increased manufacturing costs, or even weld scrapped situation. The welding sequence has a great influence on the deformation of the welding, so the study of the welding sequence has far-reaching significance in solving practical engineering problems. In this paper, the method of combining finite element with experiment is used to analyze the influence of welding sequence on the welding deformation of seal cylinder[4].

2. Surfacing model of the sealing cylinder

There are two upper and lower seal cylinders in the engine[5]. The size requirements on the sealing end are extremely strict in pairs. At the same time, there is the corrosive gas in a sealed cylinder, and we found the middle area has the outer drum phenomenon of different level in the process of casting, surfacing of thickness 1mm is required in the outer drum area. As shown in Figure 1, surfacing in the welding area on both sides of the sealed cylinder. The welding process is a local fast cooling after the heating process, which is accompanied by the production of welding deformation, the whole sealing cylinder will scrap if the welding deformation makes two upper and lower sealing face produce the error of the size, so we need to verify the effects of surfacing on seal face size.

Content from this work may be used under the terms of the Creative Commons Attribution 3.0 licence. Any further distribution of this work must maintain attribution to the author(s) and the title of the work, journal citation and DOI.
Published under licence by IOP Publishing Ltd

IOP Conf. Series: Materials Science and Engineering **361** (2018) 012018 doi:10.1088/1757-899X/361/1/012018

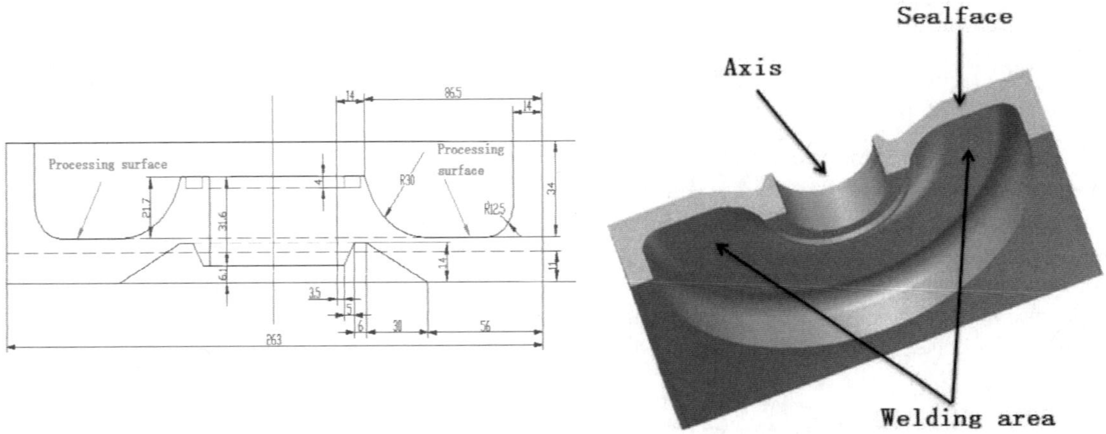

(a) The size of the sealed cylinder (b) Sealed cylinder model

Figure 1. The sealed cylinder structure

3. Finite element simulation method[6]

3.1. Finite element model

During surfacing, there is deformation and heat conduction in the surfacing area. Therefore, the surfacing area expands downward by 3mm and expands to both sides by about 10mm. The new surfacing area is the elastic zone. It is regarded as the plastic zone away from the new surfacing zone. For the accuracy of calculation, the elastic area with hexahedral mesh should be refined as much as possible. The plastic region should be tetrahedron. The connecting region between the tetrahedron and the hexahedron mesh should use Tie, so as to ensure the node correspondence. The finite element model of sealed cylinder is shown in Figure 2.

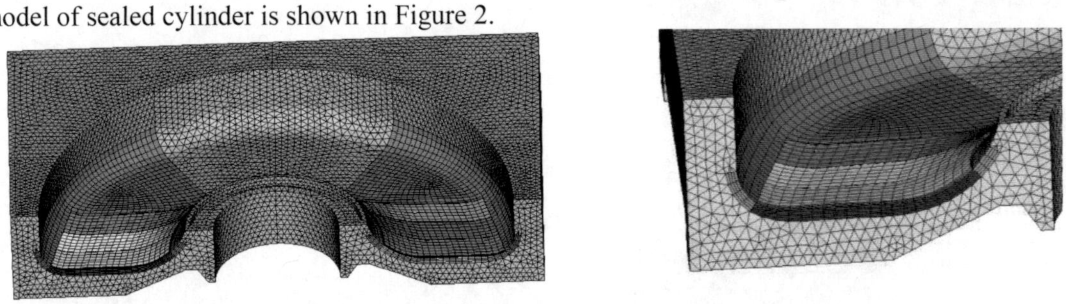

Figure 2. The finite element model of sealed cylinder

3.2. Choice of welding heat source

There are three welding heat source loading methods in welding numerical simulation, Gaussian distribution function, double ellipsoid distribution function, life and death unit method[7,8], respectively. In this study, the life-and-death unit method was used to simulate the formation of welding seam and the movement of welding heat source. Before calculation, kill all the units in the weld equivalent to the pre-weld assembly. During the calculation, the "killed" cells were "born" sequentially, simulating the filling of the weld metal. At the same time, in different beads, a corresponding heat generation rate (HGEN) is applied to the activated cell and the action time of the heat load is equal to the actual welding time. After each step of calculation, the heat rate of this step will be deleted, and the calculation is re-entered into the next step[9,10]. The heat rate is calculated as:

$$HGEN = Q(A_{weld} \times v \times dt)^{-1}$$

Where $HGEN$ is the heat rate applied by each load step, W/m^3; Q is heat, $Q = \eta UI$,which η is

IOP Conf. Series: Materials Science and Engineering **361** (2018) 012018 doi:10.1088/1757-899X/361/1/012018

arc thermal efficiency, its value is 0.7～0.9; A_{weld} is cross-sectional area of weld; v is welding speed; dt is time step for each load step, s.

3.3. Boundary conditions

For surfacing process, preheat treatment is required to meet the welding process requirements in surfacing area, so the bottom of the surfacing layer is individually selected to complete the heat treatment. The single-pass weld heating time and cooling time distribution is set to 3s and 10000s, set up a brief heat treatment process after the last weld, which insulation time is 100s, and finally cooled to room temperature in preheat treatment process. The heat transfer coefficient is set to 8, the body heat source is applied and the heat flux density is set to 1.1e7. To overcome the rigid body movement, the bottom of the sealed cylinder is fixed by using a symmetrical constraint during loading.

Heat flux density: $$q_i = \eta UIV_H^{-1}$$

Where η is arc thermal efficiency, U is Arc voltage, I is welding current, V_H is heat source volume.

4. Results of finite element analysis

4.1. The choice of welding sequence in figure 3.

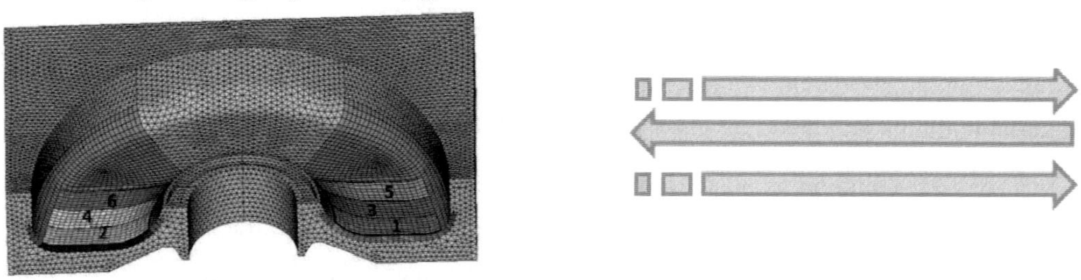

Figure 3.The welding sequence

4.2. Welding temperature field calculation results

Figures 4 (a) and (b) show the temperature distribution at the end of welding and cooling to room temperature, respectively.

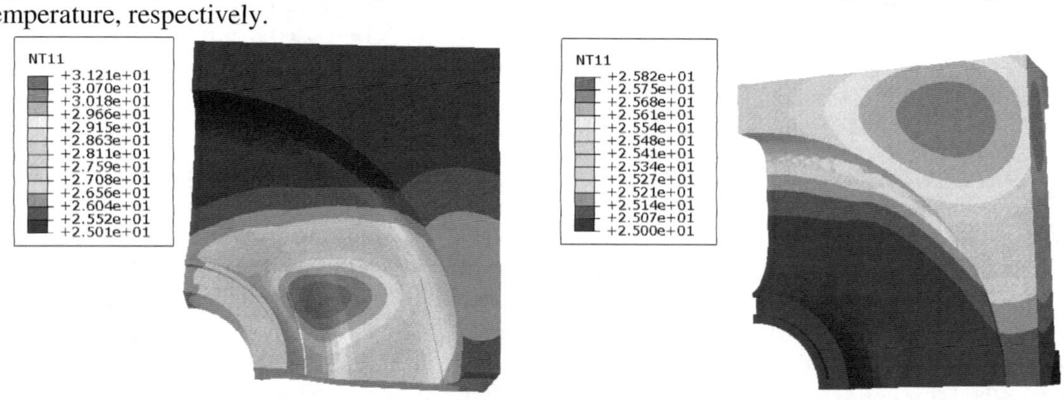

（a） Temperature distribution after the welding

（b） Temperature distribution after cooling to room temperature

Figure 4. The welding temperature distribution cloud

4.3. The calculation results of welding deformation

The deformations in XYZ directions are shown in Figure 5. Figure 5 (a) shows the amount of deformation in the X direction, which is the flatness of the work-piece after surfacing. And then by measuring the average deformation between the work-piece and the constraint surface to get the mid-face warping deformation on selecting six points in the corresponding position of the middle face. The maximum deformation of which does not exceed 0.11 mm. Figure 5 (b) is the deformation in the Y direction with the maximum deformation less than 0.13 mm. Figure 5 (c) is the deformation in the Z direction, the maximum deformation value is less than 0.16mm.

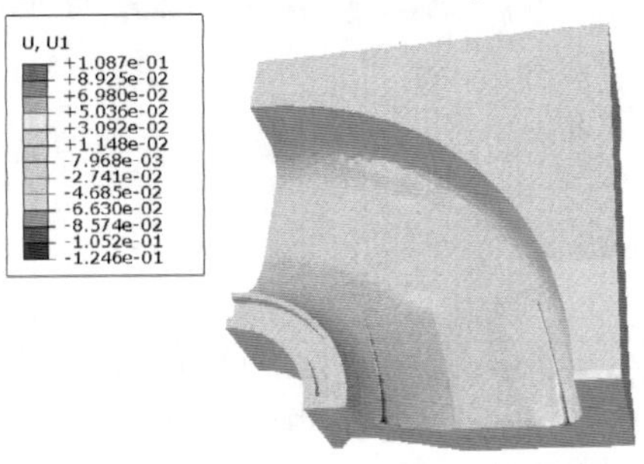

（a）The deformation in X-direction

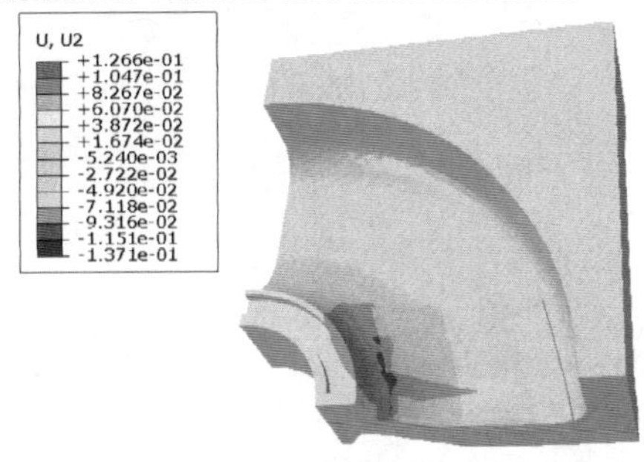

（b）The deformation in Y-direction

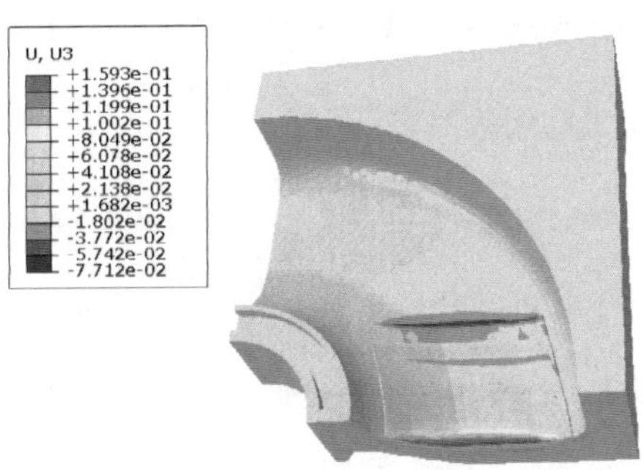

（c）The deformation in Z-direction

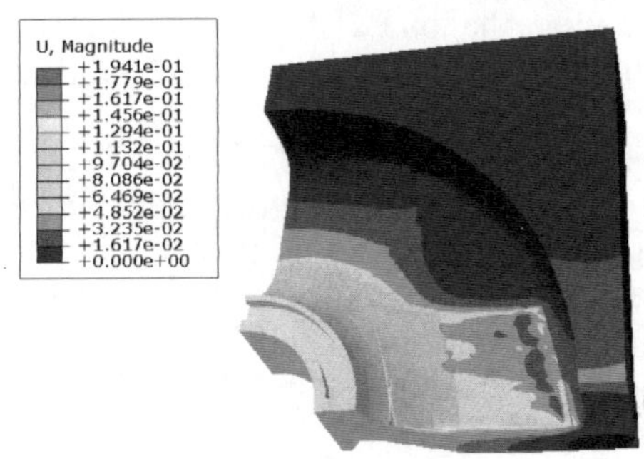

（d）Total deformation

Figure 5. Figure with surfacing deformation distribution cloud chart

The key positions in the overlay welding process include the seal face size, span size and axial seal size. The dimensions of the seal cylinder model are selected from 1-1, to 6-6, , and the cross-section is shown in Figure 6. The calculation results of deformation are shown in Table 1.

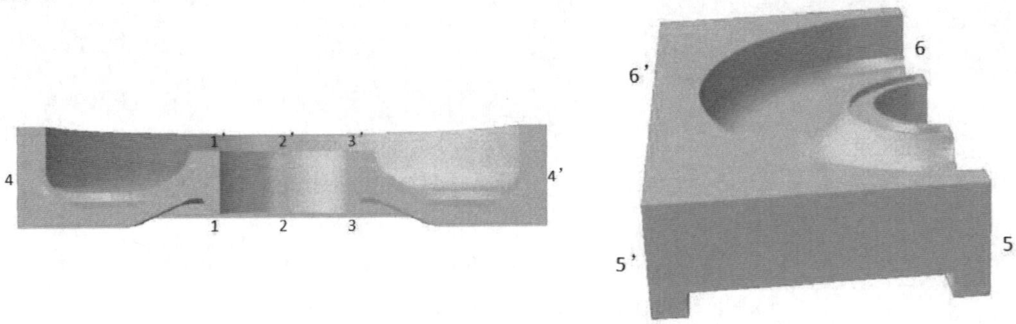

Figure 6. The key location size

IOP Conf. Series: Materials Science and Engineering **361** (2018) 012018 doi:10.1088/1757-899X/361/1/012018

Table 1. The deformation of the cross-section size

Cross-section	Direction	Before deformation (mm)	After deformation (mm)	Deformation (mm)
1-1'	Y-direction	33	32.972	-0.028
2-2'	Y-direction	33	32.956	-0.044
3-3'	Y-direction	33	32.932	-0.068
4-4'	X-direction	263	262.993	-0.007
5-5'	Z-direction	140	140.011	0.012
6-6'	Z-direction	140	140.009	0.009

The calculation results indicate the deformation in the axial direction (Y direction) is small, and the more axial deformation from the axis closer to the weld repair area through the section 1-1' , 2-2' , and 3-3' , but the maximum value is not more than 1mm; section 4-4' is the span size changes, the maximum deformation is less than 1mm. Unlike the deformation at the seal face, the farther away from the weld, the smaller the deformation appear; the cross section 5-5' and 6-6' are the deformation along the Z direction, and the deformation is smaller too.

5. Experimental results

In order to ensure the reduction of error, three models were processed. Before finishing welding, the welding position of the experimental part was determined. The MAG welding method was adopted. The welding base material selected was stainless steel and the welding material was Y316 welding wire. The welding current is 120A and the voltage is 10V. The measured position is consistent with the position of the finite element calculation model. Surfacing effect map of experimental pieces is shown in Figure 7.

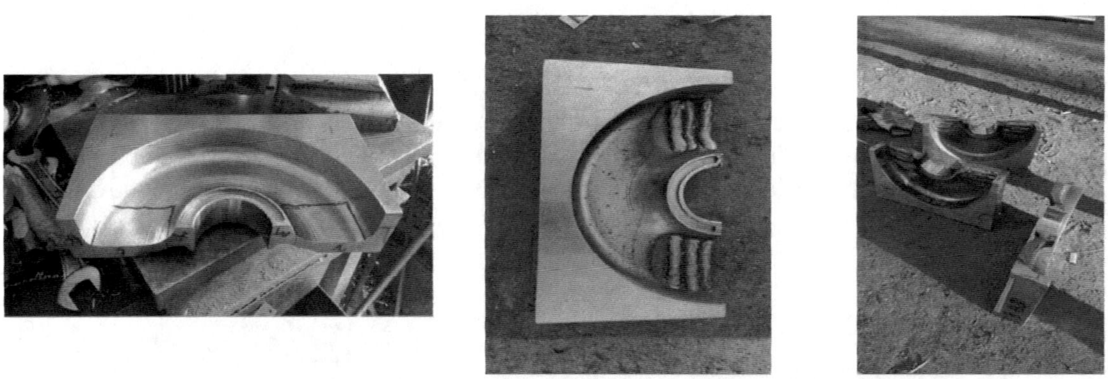

Figure 7. The surfacing effect map of experimental pieces

The experimental results in the cross-section size deformation are shown in Table 2.

Table 2. The deformation of the cross-section size

Cross-	Direction	Before deformation	After deformation	Deformation

section		（mm）	（mm）	（mm）
1-1'	Y-direction	33.02	32.94	-0.08
2-2'	Y-direction	33.027	33	-0.027
3-3'	Y-direction	33.02	33	-0.002
4-4'	X-direction	263.012	262.958	-0.054
5-5'	Z-direction	140.056	140. 072	0.012
6-6'	Z-direction	140.073	140.092	0.019

6. Conclusions

The following conclusions are shown from Table 1 and 2:

(1) Finite element calculation results and experimental calculation results have the same trend of deformation in all directions. The results obtained is more ideal in the finite element calculation, which values are less than the experimental results except for the finite element calculation cross-section 3-3' , and verify the accuracy of finite element calculation.

(2) The warping deformation of middle face in seal cylinder does not exceed the maximum 0.13mm. The sealed cylinder axial size is smaller too, and the maximum deformation is less than 0.2mm. The calculation results show that the scheme can be implemented.

References

[1] Ren Y Y and Wei S Z 2012 *Development and Prospect of Surfacing Technology (Welding Technology)* chapter 6 pp 1-5

[2] Galvis J C and Oliveira P H F 2017 *Influence of Friction Surfacing Process Parameters to Deposit AA6351-T6 over AA5052-H32 Using Conventional Milling Machine (Journal of Materials Processing Tech)* pp 91-105

[3] Huang Y X and Lv Z L 2017 *A New Method of Hybrid Friction Stir Welding Assisted by Friction Surfacing for Joining Dissimilar Ti/Al Alloy (Materials Letters)* pp 172-175

[4] Zou W H 2017 *Influence of Surfacing Reinforcement on the Deformation of the Seal Cylinder of a Steam Turbine (Welding Technology)* chapter 1 pp 76-79

[5] Bennett C 2016 *Engine Cylinder Pressure Reconstruction Using Crank Kinematics and Recurrently-trained Neural Networks (Mechanical Systems and Signal Processing)* pp 126-145

[6] Xue X L and Wang Z L 2005 *Three Dimensional Simulation of T-joing by SEM (Mechanical Engineering* vol 16*)* chapter 9 pp 811-815

[7] Chen J Q and Xiao S H 2006 *Comparison of Heat Source Mode in Numerical Simulation for Welding Process (Welding Technology* vol 35*)* chapter 1 pp 9-11

[8] Wang J 2017 *Thermal Analysis of ABS Materials based on ANSYS Life and Death Unit Technology (Synthetic Materials Aging and Application)* chapter 2 pp 41-45

[9] Chen J Q and Shen W L 2005 *Simulated Calculation of Welding Temperature based on Life and Death Element (Hot Processing Technology* vol 7*)* pp 64-65

[10] Gu J C 2014 *The Selection Principle of Heat Source in Numerical Simulation of Welding (Materials Review)* chapter 1 pp 143-146

IOP Conf. Series: Materials Science and Engineering **361** (2018) 012019 doi:10.1088/1757-899X/361/1/012019

Relative Coordinates Generality of Gradient Stability Zones and Force Factor between Hemispherical Polar Tips (in Faraday Magnetometer)

A A Sandulyak, M N Polismakova, D A Sandulyak, A V Sandulyak, D O Kiselev

Moscow Technological University, Moscow, Stromynka 20

E-mail: anna.sandulyak@mail.ru

Abstract. Coordinate dependencies of the field induction, its gradient, and the magnetic force factor in the region between hemispherical polar tips proposed for use (in the Faraday magnetometer) are given. Dependencies (and their functional form) are determined by the influence of the distance between the polar tips on the coordinates of the gradient extrema and the force factor, and also on the extreme values of these parameters: they are close to logarithmic and power-like ones, respectively. It has been shown that using hemispherical polar tips of one or another diameter, it is entirely possible to operate with more universal, namely relative (referred to the diameter or radius) values of the distance between the polar tips and coordinates of the gradient extremum or the force factor.

1. Introduction

To solve many technical and fundamental problems Faraday magnetometers of various types continue to stay in demand for a long time. In particular, they are magnetometers based on electromagnetic system use (a block consisting of magnetizing coils, cores and polar tips) [1-14], short (including superconducting) coils [15-23], permanent high-energy magnets (for example, Nd-Fe-B) [24,25].

As for the magnetometers of the first type (the most numerous), an original decision was made in [26-28] on the expediency of using hemispherical polar tips in the Faraday magnetometer. Such a solution is argued by the fact that the specially obtained coordinate (in the radial direction of the interpolar region symmetry plane) characteristic of induction B is tortuous. From a mathematical point of view, this guarantees obtaining an extreme coordinate characteristic of the induction gradient (practically identical with the partial derivative $\mathrm{d}B/\mathrm{d}x$), while in the vicinity of the x_{extr} extremum the $\mathrm{d}B/\mathrm{d}x$ values are almost stable. The coordinate characteristic of the so-called [29] magnetic force factor $B\mathrm{d}B/\mathrm{d}x$ - with the individual coordinate of the extremum x_{extr}, whose vicinity values $B\mathrm{d}B/\mathrm{d}x$ are almost stable, is also extreme.

The x_{extr} coordinates (which turned out to be independent of the current load I and dependent on the mutual removal of the hemispherical tips b) were obtained in [26-28] for the example of hemispherical tips with a diameter D = 100 mm, which, as it is, indicates their particular character. Therefore, development of researches [26-28] with use hemispherical tips of other diameter is required. In this connection, a comparative analysis of the corresponding coordinates of the stability zones of the gradient and the force factor for different diameters of the hemispherical tips D is of scientific and practical interest. In this case, it is possible to determine the eligibility of operating with relative (as more universal) parameters b/D and x_{extr}/D.

Content from this work may be used under the terms of the Creative Commons Attribution 3.0 licence. Any further distribution of this work must maintain attribution to the author(s) and the title of the work, journal citation and DOI.
Published under licence by IOP Publishing Ltd

IOP Conf. Series: Materials Science and Engineering **361** (2018) 012019 doi:10.1088/1757-899X/361/1/012019

2. Research results and their analysis

With use of polar hemispherical tips D = 135mm in diameter there are received and shown on the Figure 1 the coordinate dependencies of the field B induction in the region between the (alternatively, the dependencies for D = 100mm in [26-28]) for their various mutual distances b - from 4.7 mm to 17.6 mm and different values of the winding supply current I is from 4A to 30A. It can be seen that each of the curves B has an inflection (in the neighborhood of which, as before [26-28], its section can be linearized), which indicates the presence of an extremum of the coordinate characteristic of the gradient dB/dx (figure 2), determined, as in [26-28], on the basis of the fourth degree polynomial with the corresponding coordinates of the extremum x_{extr} (table 1). There is also a corresponding extremum of one or another coordinate dependence of the magnetic force factor BdB/dx (figure 3) - with the corresponding values of the coordinates of the extremum x_{extr} (table 2).

It should be noted that, similarly to the results obtained earlier [26-28] (using hemispherical polar tips D = 100mm in diameter), it is also characteristic here for hemispherical polar tips with a diameter D = 135mm that the coordinate of the extremum x_{extr} of both the gradient and the force factor does not depend from the current load of the winding (figure 1-3, Table 1, 2).

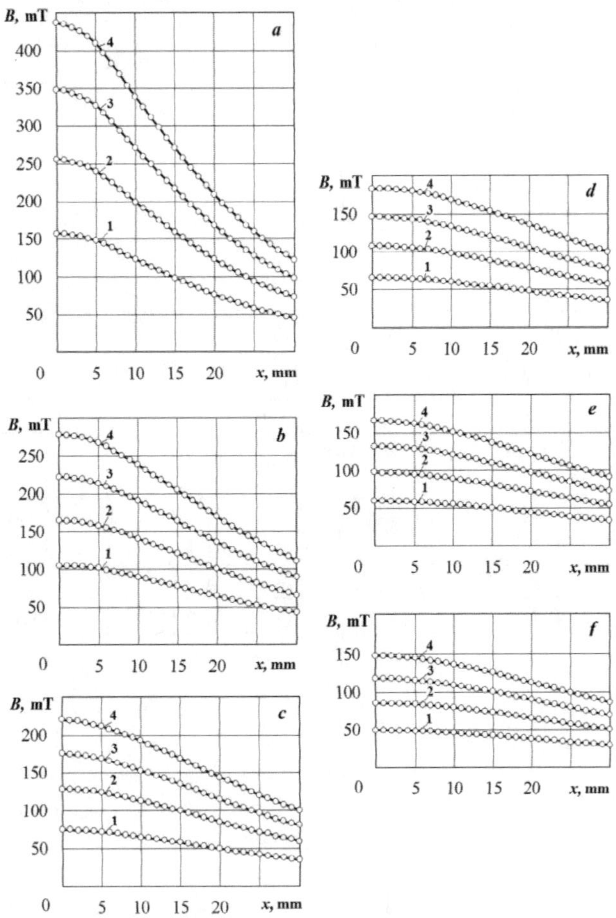

Figure 1. Coordinate characteristics of field induction between hemispherical polar tips, points - experiment, lines - calculation using a polynomial; a) b = 4.7mm, b) b = 8.1mm, c) b = 10.8mm, d) b = 13.5mm, e) b = 15.5mm, f) b = 17.6mm; 1 - I = 4A, 2 - I = 8A, 3 - I = 16A, 4 - I = 30A.

Table 1. The coordinates of the extrema of the dependencies shown in figure 2.

I, A	x_{extr}, mm					
	b=4,7mm	b=8, mm	b=10,8mm	b=13,5mm	b=15, mm	b=17, mm

4	11,43	14,62	18,23	19,86	21,57	22,94
8	11,42	14,89	16,73	19,80	21,28	19,62
16	11,43	15,07	16,71	19,55	20,72	24,93
30	11,37	14,94	16,52	21,74	20,38	21,60
	11,41	**14,88**	**17,05**	**20,24**	**20,99**	**22,27**

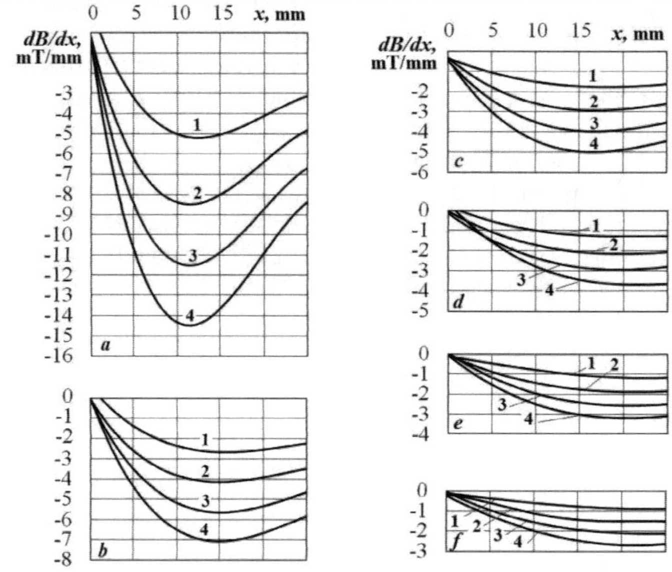

Figure 2. The coordinate characteristics of the gradient field obtained from the data in figure 1.

Table 2. The coordinates of the extrema of the dependencies shown in figure 3.

I, A	x_{extr}, mm					
	b=4,7mm	*b*=8,1mm	*b*=10,8mm	*b*=13,5mm	*b*=15,5mm	*b*=17,6mm
4	8,559	11,05	12,77	14,00	16,91	18,17
8	8,458	11,01	12,18	14,57	15,42	16,15
16	8,458	11,16	12,11	14,56	15,38	16,37
30	8,391	11,10	12,07	15,43	15,07	16,63
	8,466	**11,08**	**12,28**	**14,64**	**15,69**	**16,83**

IOP Conf. Series: Materials Science and Engineering **361** (2018) 012019 doi:10.1088/1757-899X/361/1/012019

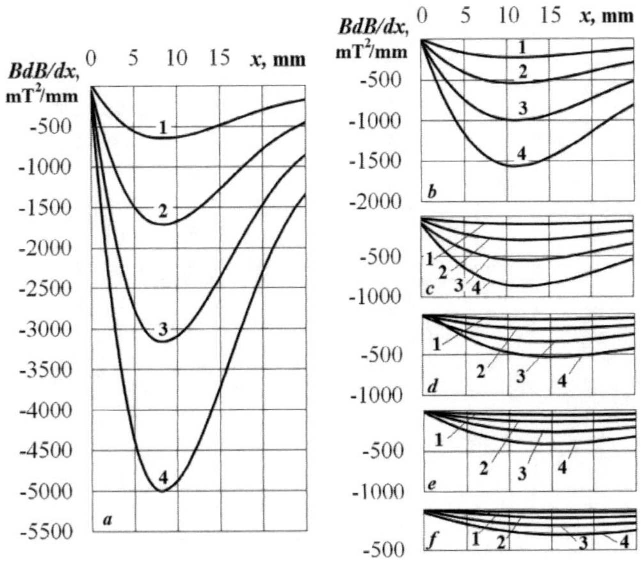

Figure 3. The coordinate characteristics of the magnetic field force factor obtained from the data in figure 1 and figure 2.

As for the influence on x_{extr} of the mutual removal b of the hemispherical tips, then, for greater values x_{extr} (in comparison with the values of x_{extr} for D = 100mm [26-28]), figure 4, the realizable data generalization in relative coordinates is noticeable: x_{extr}/D from b/D (figure 5). This testifies to the universality of the obtained results on the identification of the stability zones (gradient and force factor) in the case of using polar tips of just such a (spherical) shape.

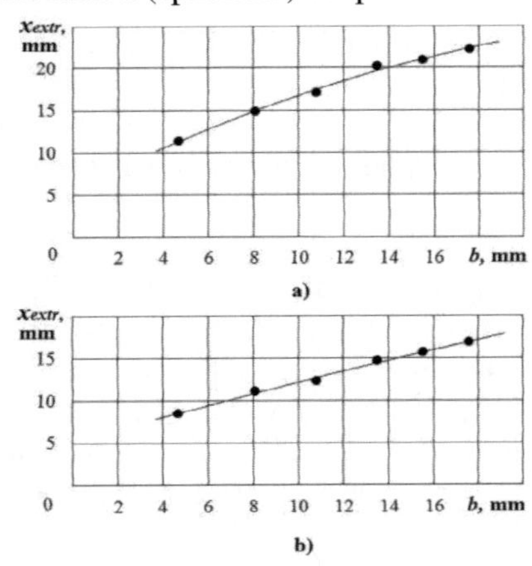

Figure 4. Influence of mutual removal of hemispherical polar tips (D = 135mm) on: a) the coordinate of the induction gradient extremum (the conditional center of its stability zone), b) coordinate of the extremum of the magnetic force factor (the conditional center of its stability zone).

IOP Conf. Series: Materials Science and Engineering **361** (2018) 012019 doi:10.1088/1757-899X/361/1/012019

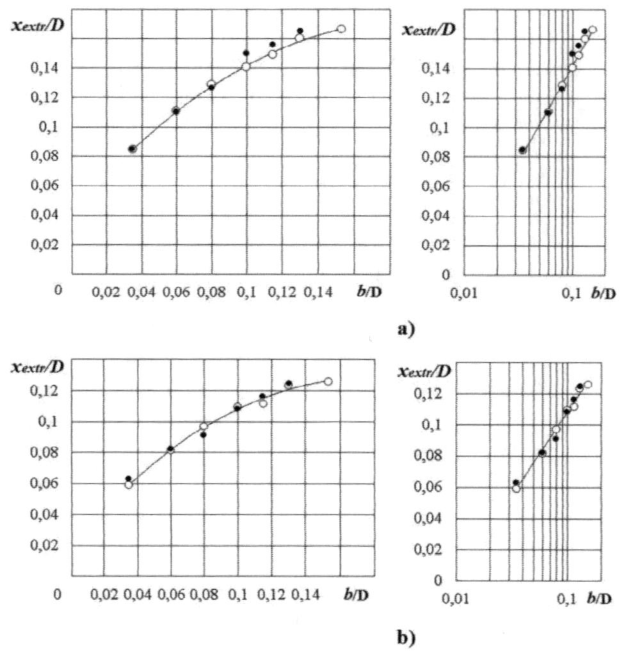

Figure 5. Influence of mutual removal of hemispherical polar tips on: a) relative coordinate of the induction gradient, b) the relative the coordinate of the extremum of the magnetic force factor; ○ - D = 100mm (data [26]), ● - D = 135mm; In semilogarithmic coordinates, the data are quasilinearizable, so that indicates their logarithmic form.

Of course, what has been said fully applies to the size of the zones themselves, i.e. Δx, Δy and Δz, the more universal representation of which is in relative form: $\Delta x/D$, $\Delta y/D$ and $\Delta z/D$ (see data in figures 1-3 in comparison with the data [26-28]). Thus, for example, if the dimensions of the zones described below are about 5x5x5mm for hemispherical polar tips of diameter D = 100mm, then these dimensions are almost 1.35 times larger, in practice, for hemispherical polar tips of diameter D = 135 mm approximately 7x7x7mm.

We also note that, judging by the quasilinearization of the x_{extr}/D data from b/D in semilogarithmic coordinates (figure 5), in the selected b/D range they correspond to the logarithmic function. In this case, the extremes of the force factor are located 30-40% closer to the axial line of the poles than for the gradient. In addition, for polar tips of hemispheres of a different diameter (in this case, D = 135 mm), as before [26-28] (D = 100mm), the dependence of the induction gradient abscissas is practically the same (figure 6a, exponent: -1, 3) and the force factor (figure 6b, exponent: -2.2) from the mutual removal of hemispherical polar tips.

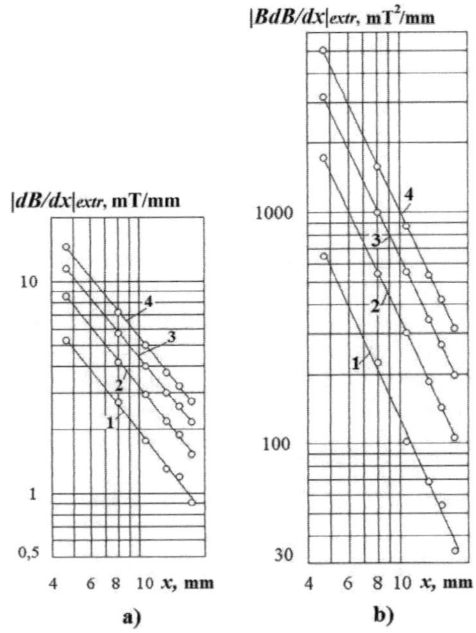

Figure 6. Illustration of the power-law dependence of the induction gradient (a) and magnetic force factor (b) from the mutual removal of hemispherical polar tips (after the quasi-linearization of these data in logarithmic coordinates); 1 - I = 4A, 2 - I = 8A, 3 - I = 16A, 4 - I = 30A.

3. Conclusion

The families of the coordinate characteristics B, dB/dx, BdB/dx are obtained and analytically analyzed, depending on the basic parameters of the unit functioning responsible for creating a non-uniform field in the Faraday magnetometer: current load I, mutual removal b between the hemispherical tips proposed for use, their diameter D. The x_{extr} extremum coordinates for the parameters dB/dx and BdB/dx are found and compared (including on the basis of the calculated dependence obtained). It is shown that for each of the values of b, the x_{extr} data for both the gradient and the force factor remain practically unchanged irrespective of I, while the extrema of the force factor are 30-40% closer to the axial line of the poles than to the gradient. The dependencies (and their functional form) are determined by the influence of b on the x_{extr} values for dB/dx and BdB/dx, and on the extreme values of these parameters: they turned out to be close to the logarithmic and power-law dependencies, respectively. It is shown that using hemispherical polar tips it is quite possible to operate with more universal ones, namely the relative parameters b/D and x_{extr}/D.

References

[1] Garber M, Henry W G and Hoeve H G A magnetic susceptibility balance and the temperature dependence of the magnetic susceptibility of copper, silver, and gold, 295^0– 975^0 K 1960 *Can J Phys* 38 1595-1613.

[2] Klaase JCP The Faraday balance, Van der Waals- Zeeman Institute, November 1999, URL: http://www.science.uva.nl/research/cmp/klaasse/fdb.html, as of December 2014.

[3] Quinn R K and Knauer R C Low temperature Faraday susceptibility apparatus 1972 *Rev Sci Instrum* 43 1543-1544.

[4] Petersson L and Ehrenberg A Highly sensitive Faraday balance for magnetic susceptibility studies of dilute protein solutions 1985 *Rev Sci Instrum* 56 575-580.

[5] Lewis R T A Faraday type magnetometer with an adjustable field independent gradient. Review of Scientific Instruments 1971 *Rev Sci Instrum* 42 31-34.

[6] Schäfer H L, Morrow J C and Smith H M Magnetic susceptibility and crystal structure of

(Pyridine *N*-Oxide)-Copper(II) Chloride 1965 *J Chem Phys* 42 504-508.

[7] Gruber J B and Hecht H G Low temperature magnetic susceptibility of UCl_4 1974 *J Chem Phys* 60 1352-1354.

[8] Gopalakrishnan R, Barathan S, Govindarajan D Magnetic susceptibility measurements on fly ash admixtured cement hydrated with groundwater and seawater 2012 *AmerJMaterials Science* 2(1) 32-36.

[9] Hennig G R and McClelland J D Magnetic susceptibility and free energy of graphite bromide 1955 *J Chem Phys* 23 1431-1435.

[10] Heyding R D, Taylor J B and Hair M L Four-inch shaped pole caps for susceptibility measurements by the Curie method 1961 *Rev Sci Insrum* 32 161-163.

[11] Hosu B G, Jakab K, Bánki P, Tóth F I and Forgacs G Magnetic tweezers for intracellular applications 2003 *Rev Sci Instrum* 74 4158-4163.

[12] Marcon P, Ostanina K March 27-30, 2012 PIERS Proceedings, Malaysia, Kuala Lumpur 420–424.

[13] Govindarajan D and Gopalakrishnan R Magnetic susceptibility measurements on metakaolin admixtured cement hydrated with ground water and sea water 2009 *International Journal of Minerals, Metallurgy and Materials* 16 349-354.

[14] Burmester W L and Sellmyer D J Are VPd_3 and $NbPd_3$ itinerant ferromagnets? 1982 *J Appl Phys* 53 2024-2026.

[15] Zhang C P, Chaud X, Beaugnon E and Zhou L Crystalline phase transition information induced by high temperature susceptibility transformations in bulk PMP-YBCO superconductor growth in-situ 2015 *Physica C* 508 25-30.

[16] Seidov Z, H.-A. Krug von Nidda, Hemberger J, Loidl A, Sultanov G, Kerimova E and Panfilov A Magnetic susceptibility and ESR study of the covalent-chain antiferromagnets $TlFeS_2$ and $TlFeSe_2$ *Physical Review B.* 65 p 014433.

[17] Slobinsky D, Borzi R, Mackenzie A, Grigera S Fast sweep-rate plastic Faraday force magnetometer with simultaneous sample temperature measurement. 2012 *ARevSciInstrum* 83 125104.

[18] Mexner W and Heinemann K An improved method for relaxation measurements using a Faraday balance 1993 *Rev Sci Instrum* 64 (11) 3336-3337.

[19] Blach T P, E MacA Gray A Faraday magnetometer for studying interstitially modified ferromagnets 1994 *Meas Sci Technol* 5 1221-1225.

[20] Reutzel S and Herlach D M Measuring magnetic susceptibility of undercooled co-based alloys with a Faraday balance 2001 *Adv Eng Mat* 3 65–67.

[21] Srinivasan R and Usha S Auxiliary coils for generating magnetic field gradients for a Faraday magnetometer 1986 *J Physics E: Sci Instrum* 19 930–932.

[22] Stewart A M The superconducting Faraday magnetometer: error forces and lateral stability 1975 *J Physics E: Sci Instrum* 8 55–59.

[23] Cape J A and Young R A Canted Helmholtz coils for constant-gradient Faraday balance magnetometry 1971 *Rev Sci Instrum* 42 1061–1063.

[24] Riminucci A, Uhlarz M, R De Santis and Herrmannsdörfer T Analytical balance-based Faraday magnetometer 2017 *J Appl Phys* 121 094701.

[25] Finot E., Thundat T., Lesniewska E., Goudonnet J P Measuring magnetic susceptibilities of nanogram quantities of materials using microcantilevers 2001 *Ultramicroscopy* 86 175–180.

[26] Sandulyak A V, Sandulyak A A, Polismakova M, Ershova V, Sandulyak D, Kiselev D On the issue of choosing the measuring zones in a Faraday balance when studying magnetic susceptibility of small samples 2016 *Applied Physics, System Science and Computers* 77-83.

[27] Sandulyak A V, Sandulyak A A, Polismakova M N, Kiselev D O, Sandulyak D A, Ershova V A The working zone in the interpolar area of the Faraday balance: an approach to testing the magnetic force factor stability criterion 2017 *MATEC Web of Conferences.*

[28] Sandulyak A A, Sandulyak A V, Polismakova M N, Kiselev D O, Sandulyak D A An approach for choosing positioning of small volume sample at instantiation ponderomotive Faraday method in determining its magnetic susceptibility 2017 *Russian Technological Journal* No. 2 57–69.

[29] A. V. Sandulyak *Purification of liquids in magnetic field* Lvov: High School 1984 p 167.

Acknowledgments
The research is conducted with financial support from RFFI within the frameworks of research project № 16-38-60034 mol_a_dk and from Russian Federation Ministry of Education and Science №9.9626.2017.

IOP Conf. Series: Materials Science and Engineering **361** (2018) 012020 doi:10.1088/1757-899X/361/1/012020

Mathematical Modeling of Mechatronics System "Hydraulic Screwdown Mechanism - Electric Drive of Rolling Mill Stand"

A A Radionov[1], V R Gasiyarov[1], S N Baskov[1], A S Karandaev[1], V R Khramshin[2]

[1]South Ural State University, 76, Lenin Avenue, Chelyabinsk 454080, The Russian Federation
[2]Nosov Magnitogorsk State Technical University, 38, Lenin Av., Magnitogorsk, 455000, The Russian Federation

E-mail: gasiyarovvr@gmail.com

Abstract. The article provides a rationale for the mathematical model of a mechatronics system "hydraulic screwdown mechanism - electric drive of a rolling mill stand". The authors emphasized the relevance of the research of the developed method to limit dynamic loads imposed on electromechanical system rolls in compliance with which a roll gap is increased before metal bite and roll movement to a set position during rolling. The paper presents a block diagram for a complex mathematical model including local models of a hydraulic screwdown mechanism in the automatic gage control system (AGCS) and a rolling mill stand structure interconnected by metal. Further, the setup of a system for the adjustment of the screwdown mechanism position is considered. The authors provide analytical dependences to calculate the specific roll force and mill torque. The developed method was modeled, and it was proved that its implementation results in the decrease of the dynamic torque at metal bite. The article is concluded with the recommendations on the developed model use.

1. Introduction

Plate rolling mill operation is associated with the problems of the limitation of dynamic loads when metal is bitten with the horizontal rolls of a reversing rolling stand. Such problem is characteristic of the plate rolling mill 5000 of Magnitogorsk Iron&Steel Works. The main technological device of a rolling mill is a four-high mill stand providing for a roll force up to 12,000 tons. It is the most high-duty stand among all the similar mechanisms in Europe.

The solution of the aforementioned problem is complicated by the fact that the software of the stand automatic process control system is not accessible for the settings adjustment. It limits the possibilities to conduct active experiments at the rolling mill. That is why the most efficient way to conduct the research is mathematical modeling.

A mathematical model to be developed shall have a universal nature. It shall include the data from a mechatronics system of any sheet or wide-strip mill. But the model structure shall remain the same. The modeling ultimate objective is to develop recommendations to introduce the methods aimed at the reduction of dynamic loads on the rolling mill 5000.

2. Task statement

Content from this work may be used under the terms of the Creative Commons Attribution 3.0 licence. Any further distribution of this work must maintain attribution to the author(s) and the title of the work, journal citation and DOI.
Published under licence by IOP Publishing Ltd

IOP Conf. Series: Materials Science and Engineering **361** (2018) 012020 doi:10.1088/1757-899X/361/1/012020

The method developed for the reduction of dynamic loads applied to the plate rolling mill provides for an automatic gap increase before metal bite. After some pause following the metal bite a hydraulic screwdown mechanism moves into a set position determined by the reduction value. To study this method, it is necessary to develop a mathematical model for a mechatronics system "hydraulic screwdown mechanism - electric drive of a rolling mill stand in the AGCS structure" taking into account the interconnection of these mechanisms by metal. One shall also take into consideration that modeling does not cover the research of "internal" coordinates of the electric rives (currents, voltages, magnetic flow, etc.). That is why the electric rive model shall be taken as a simplified one without taking into account the peculiarities of synchronous motor vector control [1].

3. Body

The stand of the plate rolling mill 5000 is equipped with two working rolls transmitting the rolling effort and electric drive power directly to the metal being rolled. The deformation zone in the stand is formed from the top with the mechanic screwdown mechanism and from the bottom - with the hydraulic screwdown mechanism [2]. The electric drive of each roll is individual. The mechanic joints do not have the mechanisms with the inertia (mass) comparable to the motor inertia and corrected inertia of the working and support rolls. That is why the electromechanical system is adopted as a dual-mass one with the elastic constraint and gap in mechanical joints. The model block diagram is presented in Figure 1. The block 2 determines the nature of transition processes in the mechanical part including natural oscillations damping, the block 3 models the gaps in mechanical transmissions while the blocks 1,4,5 are the standard blocks of a dual-mass model of electromechanical systems [3].

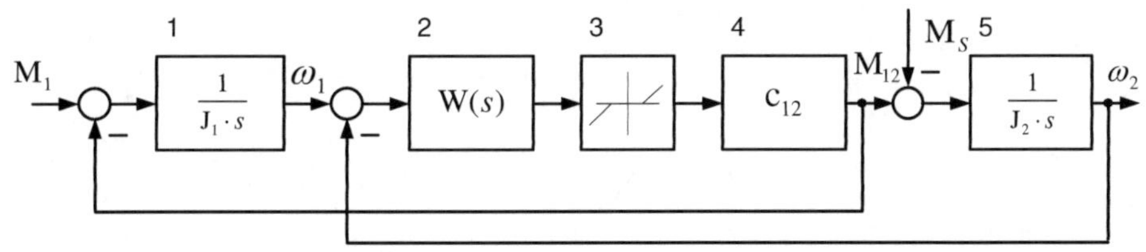

Figure 1. Block diagram of the roll electric drive dual-mass system.

Such structure is a component of a comprehensive model of the studied mechatronics system the diagram of which is given in Figure 2. The electric drive control system is a dual-circuit system with an external circuit of speed control. The internal closed loop for current control is represented by the aperiodic first-order factor. The structure also includes the hydraulic screwdown mechanism model as a part of the AGCS and the model of interconnection between hydraulic and electromechanical systems by metal which is a simplified deformation zone model in terms of its nature. They are built on the basis of known analytical dependencies considered below.

Mathematical model of hydraulic screwdown mechanisms in the AGCS structure

The conducted analysis of literature references showed that there are several variants of the hydraulic screwdown mechanism description as a control object [4-6]. Below are the main dependences to build a model reasoned in detail in [7,8]. The screwdown mechanism as a control object is described by three linearized differential equations.

1. The fluid flow rate equation for coordination small increments:

$$Q = S_G \frac{dS_{sd}}{dt} + \frac{V_{sd}}{E}\frac{dP_G}{dt} + rP_G \qquad (1)$$

where Q – rate of the fluid flowing into a hydraulic cylinder piston cavity; S_{sd}, S_G – position and area of the hydraulic cylinder piston; V_{sd} – volume of the piston cavity in the piston initial position; E – volumetric modulus of elasticity of the hydraulic fluid; P_G – pressure inside the piston cavity; r – leakage coefficient.

IOP Conf. Series: Materials Science and Engineering **361** (2018) 012020 doi:10.1088/1757-899X/361/1/012020

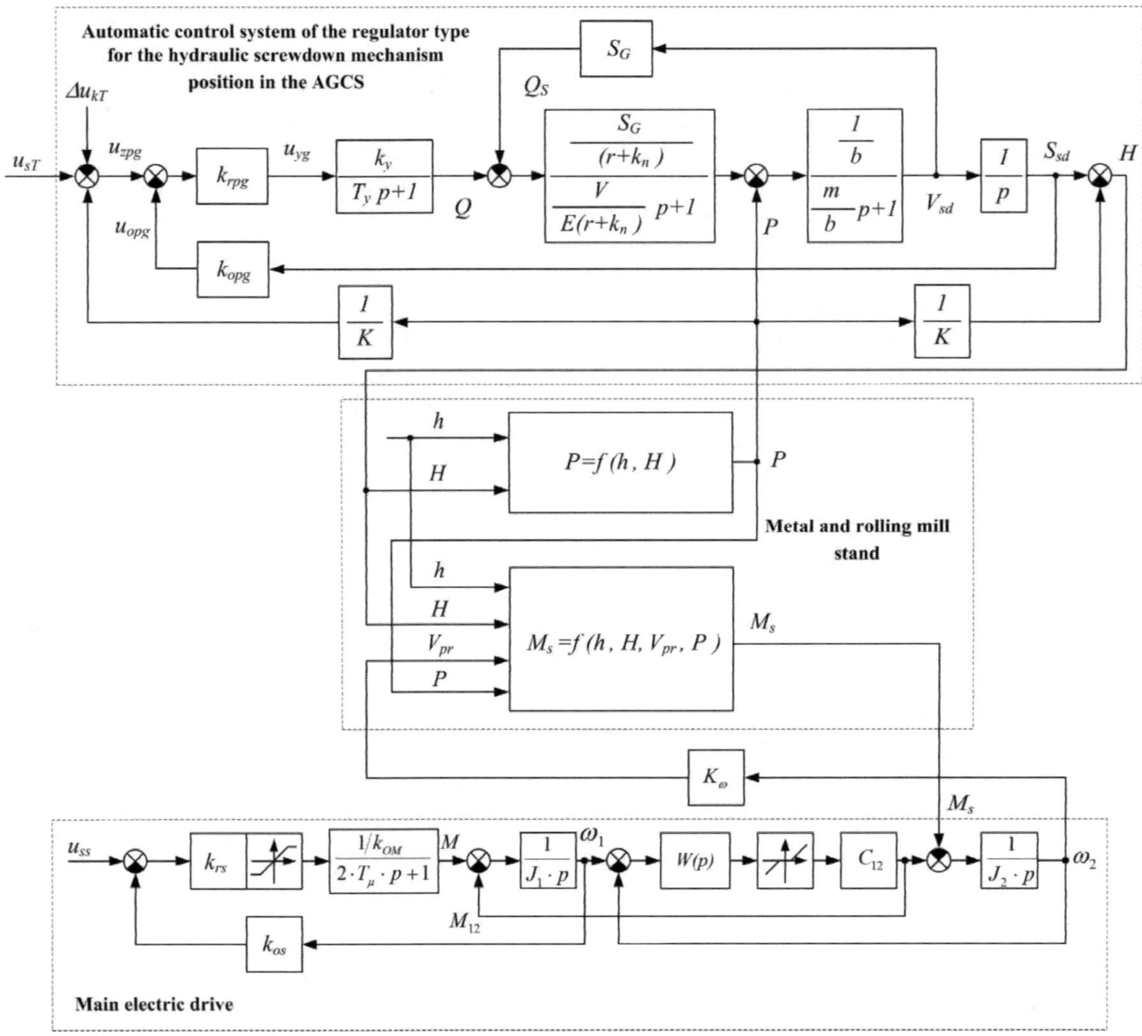

Figure 2. Block diagram of the developed mathematical model.

2. The equation describing the effort ratio inside the stand:

$$P_G S_G = P + m\frac{d^2 S_{sd}}{dt^2} + b\frac{dS_{sd}}{dt} \tag{2}$$

where P – full roll force; m –mass of the stand mechanical parts moving together with a hydraulic cylinder as well as the mass of the fluid in the pipeline running from the battery to the hydraulic cylinder reduced to the piston area; b – viscous friction coefficient determining the value of the friction force proportionate to the hydraulic screwdown mechanism travel speed.

3. The linearized equation of the servo-motored valve for coordinate small increments:

$$T_y\frac{dQ}{dt} + Q = k_y u_y - k_{pr}P \tag{3}$$

where T_y – servo-motored valve time constant; u_y – control voltage; $k_y = \frac{dQ}{du_y}$ – constant flow rate coefficient of the servo-mored valve; $k_{pr} = \left(\frac{dQ}{dm}\right)_0$ – stiffness coefficient of difference flow characteristics at the pressure (P_0) inside the piston cavity taken as the initial one.

The given equations (1)-(3) do not take into account:

- wave processes in the pipeline connecting the battery, servo-motored valve and hydraulic cylinder;
- pressure loss in the pipeline;
- dry friction.

The transmitting function of the screwdown mechanism position adjuster at the technical optimum setting [9]:

$$W_{prg}(p) = \frac{1}{W_{oy}(p)\, k_y\, k_{opg}} \frac{1}{2T_y p}$$

where $k_{opg} = \frac{U_{sp\ max}}{S_{sd\ max}}$ – feedback coefficient on the hydraulic cylinder piston position; $W_{oy}(p)$ – transmitting function of the hydraulic screwdown mechanism as a control object.

If we ignore the high-order values in the transmitting control function, we can apply a proportional screwdown mechanism position adjuster with the amplification coefficient κ_{rpg}. Figure 3a, presents logarithmic amplitude frequency characteristics (LAFC) of the open and closed circuits of the position adjuster at technical optimum setting. They are built for the proportional adjuster with the coefficient $\kappa_{rpg} = 95$. The curves of the transition process for the screwdown mechanism position at the change of the task with κ_{rpg}=95 are given in Figure 3b.

At such setting, the screwdown mechanism position re-adjustment is equal to 30%. At the same time, in compliance with [10], the re-adjustment value shall not exceed 5%. To fulfill this condition, the amplification coefficient is taken as $k_{prg} = 65$. The corresponding LAFC of the open and closed circuits of the position adjustment and the transition process graph are also given in Figure 3a and Figure 3b. In this case, the setting provides for the fulfillment of the set requirement.

Hydraulic screwdown mechanisms control is effected with the help of an indirect system regulating the sheet (plate) gage built on the basis of the Golovin-Sims equation [11]:

$$\mathrm{H} = d + \frac{P}{M_K}$$

where H – thickness of the plate at the stand outlet; d – roll gap size; P – roll force; M_K – stand stiffness modulus.

Based upon this expression, the AGCS block diagram is built in the model in Figure 2.

Mathematical model of rolling in the deformation zone

At the mathematical model development the authors assumed the following:

- there is no flare at rolling;
- initial rolling gage at the stand inlet does not change within the analyzed time period;
- distribution of the metal delivery rate is equal along the whole section depth.

IOP Conf. Series: Materials Science and Engineering **361** (2018) 012020 doi:10.1088/1757-899X/361/1/012020

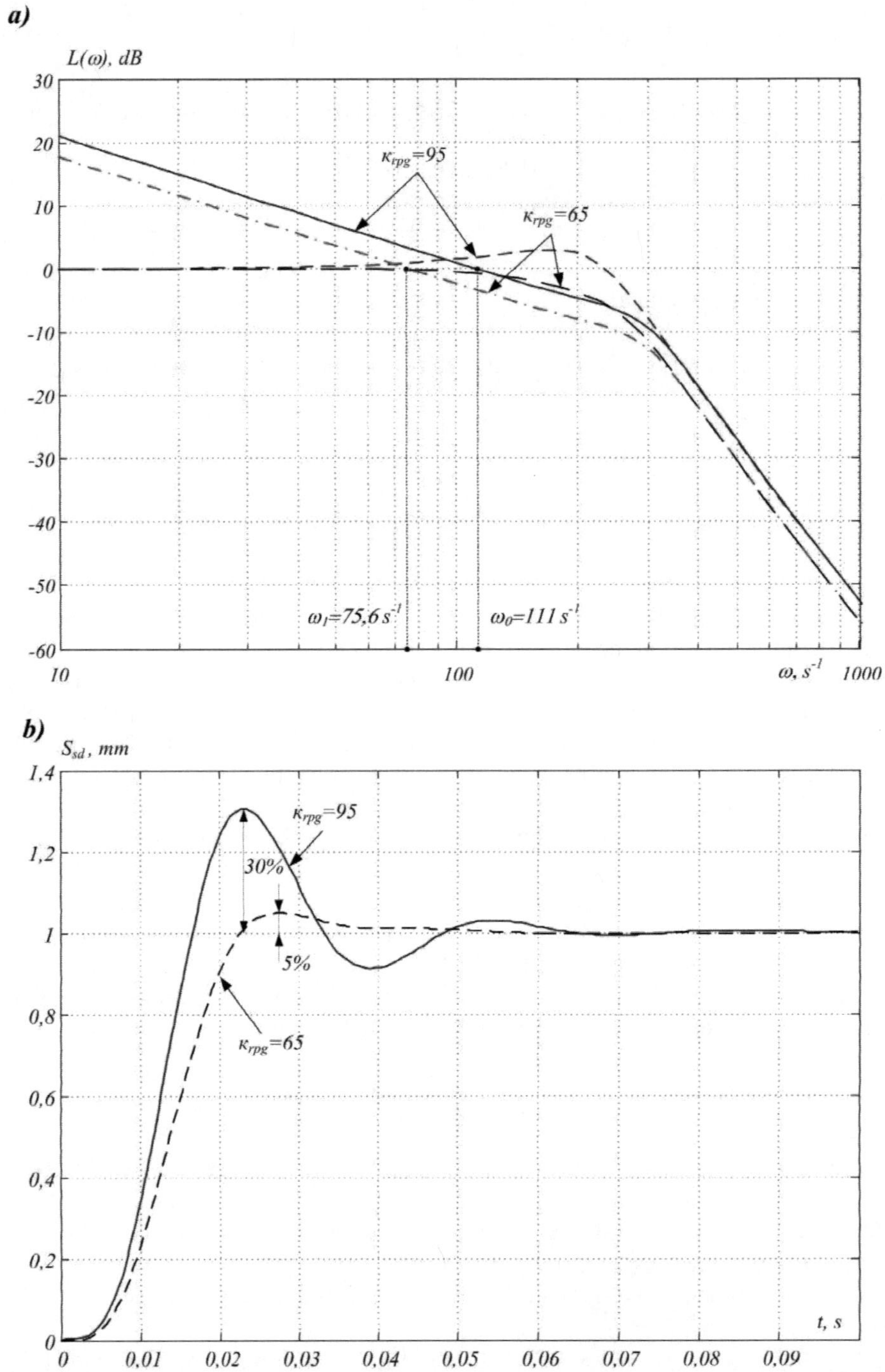

Figure 3. LAFC of the open and closed circuits of the scredown mechanism position adjustment (a) and the graph of the transition process at the scredown mechanism position change (b) at $k_{prg} = 95$ and at $k_{prg} = 65$

The rolled sheet as a control object is described by the known equations. To determine the forward flow at rolling, the authors use the Dresden formula [12]. To calculate the roll force, the authors use the expression proposed by A.I. Tselikov [13]:

$$P = \frac{B \cdot X \cdot H}{2\mu}\left[\frac{1}{\phi}\left(\frac{X_{i-1}}{X_i}\right)^{\phi} e^m - \left(\frac{X_{i-1}}{X_i} - 1\right) - \frac{1}{\phi}\right] \tag{4}$$

where B – sheet gage; $\phi = \frac{h}{h+H}$; $m = \frac{2\mu \cdot l}{h+H}$; $X_{i-1} = 1{,}15\sigma_{s\,i-1} - \sigma_{i-1}$; $X_i = 1{,}15\sigma_{s\,i} - \sigma_i$ – coefficients; σs i-1, σs i – metal yield strength before and after reduction in the stand (at loose rolling inside a reversing rolling stand these values are taken as equal ones, that is why they are not taken into account in Figure 2 [14,15]); μ - friction coefficient between the working surface of the rolls and metal; $l = \sqrt{R(h-H)}$ – length of the arc of roll contact with the account of roll flattening.

The static torque on the motor shaft is determined as a total of components:

$$M_{ss} = M_{rol} + M_{fr1} + M_{fr2} \tag{5}$$

where $M_{rol} = 2 \cdot P \cdot \psi \cdot 1$– mill torque determined in compliance with the roll force; $M_{fr1} = P \cdot D \cdot \mu_1$– torque of the additional friction forces; $M_{fr2} = \left(\frac{1}{\eta} - 1\right) \cdot \frac{M_{rol} - M_{fr1}}{i}$ – friction torque in transmission equipment.

In these expressions: R – roll radius; ψ – rolling arm coefficient; D – roll neck diameter; μ_1 – friction coefficient for film-lubrication bearings; η – transmission efficiency coefficient; i – gear reduction rate.

The model block diagram in Figure 2 is made with the use of the presented mathematical apparatus. In the model structure the dependences (4), (5) are implemented in the functional blocks *P=f(h, H)* and $M_s=f(h, H, V_{pr}, P)$.

Obviously, at the assumptions made, one shall pay attention to the assessment of the model correlation with the studied object. Such assessment was conducted for the whole mechatronics complex as well as individually for the main electric drive and screwdown mechanism hydraulic drive. The analysis was carried out by means of comparison of the modeling results with the data obtained by oscillography at the mill 5000. The authors analyzed the transmission processes of the electric and hydraulic drive coordinates at rolling according to the "traditional" pattern and in compliance with the proposed method. The results of the comparison made for the nature of the processes as well as for the estimated and experimental values in the fixed points allowed the authors to make a conclusion on their mutual correspondence. As a result, the model-object correlation was proved with a sufficient degree of accuracy.

4. Modeling results

The developed model was used for the research of the method proposed for dynamic loads limiting. In compliance with the research programme, the gap before metal bite was made larger than it is required to obtain a target gage and was further decreased to the set value as the head part enters the stand. It was assumed that the load torque at metal bite changes abruptly.

The Figure 4 gives time dependences for the electric drive speed and torque change. Metal is bitten at the torque of 2.25 c with the increase of the roll gap and at different initial reductions: curves 1 – at 70%, curves 2 – at 80% and curves 3 – at 90% of nominal reduction. In 0.25 c after that, within 2.5–3 sec. the linear gap decrease occurs due to the hydraulic screwdown mechanisms movement. Accordingly, under the linear law the static load torque increases (window 2). Meanwhile, the mill

IOP Conf. Series: Materials Science and Engineering **361** (2018) 012020 doi:10.1088/1757-899X/361/1/012020

torque reaches the set value while the gap reaches its set size corresponding to the sheet gage after one pass.

The analysis of the presented graphs demonstrated that in case of metal bite at the preliminary roll convergence at 90% of the required roll gap, the torque re-adjustment (curve 3) is equal to 50%. At the graphs 1 and 2 the torque re-adjustment is equal to 13% and 27%, correspondingly. Consequently, the effect of torque fluctuation limiting depends on the initial roll setting.

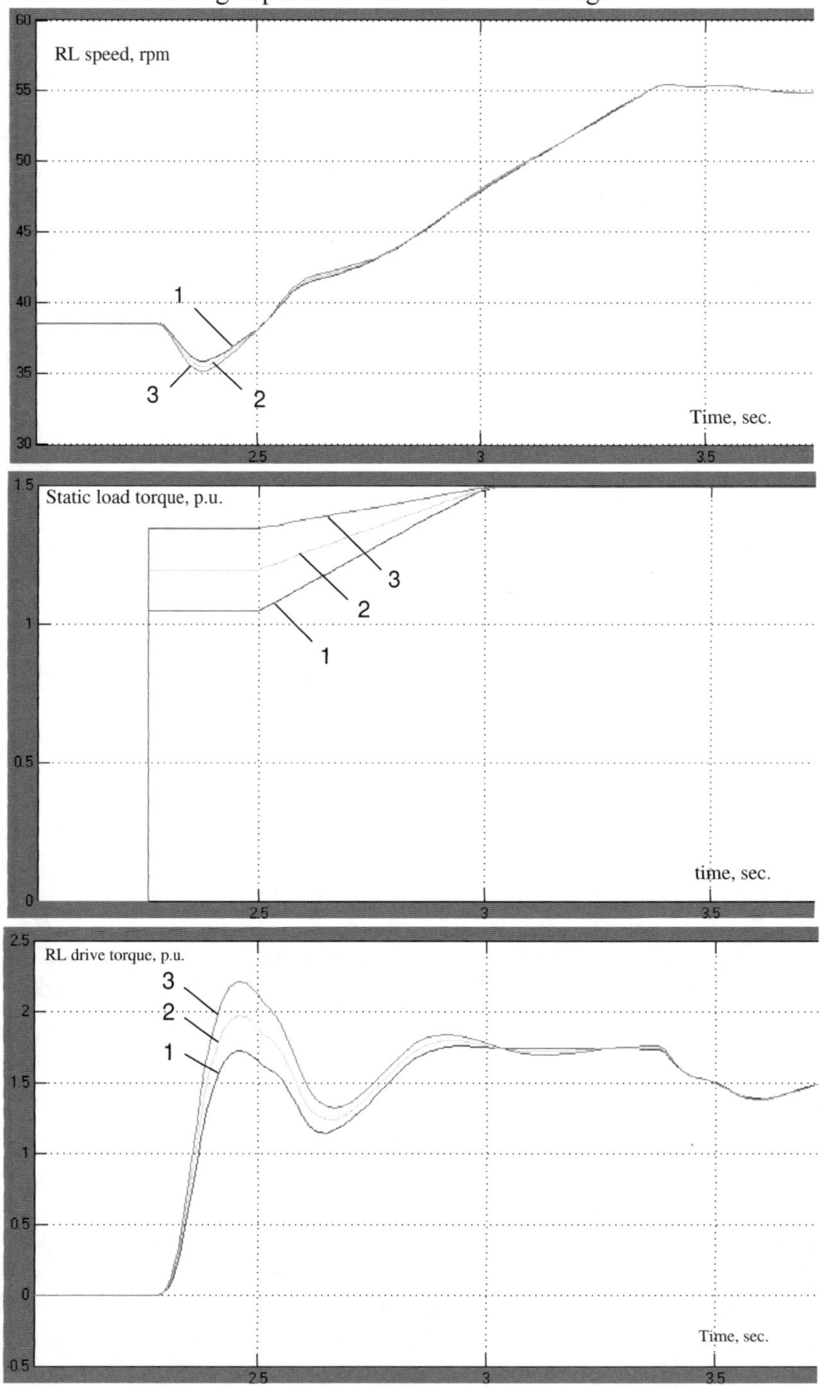

Figure 4. Speed and dynamic torque transition processes for various gaps at metal bite.

5. Discussion

The modeling results confirm that the proposed method implementation provides for the dynamic torque decrease. The limitation is the roll gap size set before metal bite. The effective torque decrease can be reached if the roll set within the 70% of the set reduction (curve 2 in Figure 4). However, under such conditions it is necessary to significantly move the rolls under the load; such movement can cause section deformation in the rolling head part. Besides, at heavy reduction the load onto a hydraulic screwdown mechanism will significantly increase.

It is clear that this method application at the rolling mill 5000 will not always provide for the expected results. It is explained by a number of reasons, in particular, by the fact that a deformation zone filling at metal bite does not occur immediately as it was assumed at modeling but takes several seconds. That is why roll force changes in accordance with a complex law, therefore, the transition processes will be different from the presented ones. In this connection, it is necessary to conduct additional research using both mathematical modeling and direct studies on the object.

6. Conclusion

As a result of the conducted studies the authors obtained a relatively simple mathematical model of a mechatronics system "electric drive of a rolling mill stand-hydraulic screwdown mechanism in the AGCS structure" containing the models of the following devices and systems:

- dual-mass electromechanical system of the main electric drive line;
- hydraulic screwdown mechanism;
- systems adjusting the screwdown mechanism position with a proportional position adjuster;
- systems of gage indirect control based on the Golovin-Sims equations;
- interconnection of electric and hydraulic drives by the treated metal.

The model is implemented as a programme in the graphical environment of imitation modeling of dynamic systems Simulink. In addition, the authors developed a special programme to calculate roll force and mill torque dependences. It allows for the calculation of mathematical model coefficients for various rolling modes applied at the mill in practice.

The authors modeled the developed method for dynamic torque limitation due to the increase of the roll gap before metal bite with its subsequent return into the set position during the rolling process. It was demonstrated that the proposed method provides a solution for the set tasks under the ideal bite conditions. However, its practical application under the conditions of the rolling mill operation needs to be further investigated.

The as-is model is suitable for the research of any modes associated with roll gap change during rolling, in particular to study the rolling process with contouring of wide rolling margins by horizontal rolls [16]. After proper follow-up revision such model may be applied to study the modes of mechatronics systems in hot continuous rolling mills [17]. For this purpose it shall be completed with a dynamic plate model in the interstad space [18,19].

References

[1] Shreiner R T 2000 *Mathematical Modelling AC-Drives with Semiconducting Frequency Convertor* (Ekaterinburg: Ural Department of Russian Academy of Sciences) p 654

[2] Shilyaev P V, Usaty D Yu and Radionov A A 2011 Automation of Plate Rolling at 5,000 mm Mill *Russian Electromechanics* **4** pp 15–8

[3] Gasiyarov V R and Maklakova E A 2015 Mathematical Description of Main Electric Drive of Hot Plate Mill 5000 *Russian Internet Journal of Electrical Engineering* vol 2 **3** pp 62–6

[4] Shaw D A, Foulds J G and Horner A C 1988 Custom design of hydraulic gauge control for three Canadian hot strip mills *Iron and Steel Eng* **12** pp 21–9

[5] Kuzishcin V A and Gedimin Yu Yu 1985 *Modeling of hydraulic pressing devices of rolling mills, proceedings of Designing and research of modern rolling mills* (Moscow, VNIImetmash) pp 113–8

[6] Lederer A 1982 State of development of plate mills *MPT* **5** pp 36–60

[7] Braun A E, Draluk B N and Troitskiy A E 1982 Some questions of the dynamics of hydraulic jacks *Electrotechnical industry. Ser. Electric drive* vol 8 pp 13–7

[8] Hramshin V R, Karandaev A S, Radionov A A and Hramshin R R 2013 Study of Thickness Control of Strip Head Section Using Mathematical Simulation Methods *Herald of SUSU. Series Energy* is 13 **1** pp 144–51

[9] Karandaev A S, Radionov A A, Khramshin V R, Andryushin I Yu and Shubin A G 2014 Automatic Gauge Control System with Combined Control of the Screw-Down Arrangement Position *12th Int Conf on Actual Problems of Electronic Insrument Engineering (APEIE-2014)* vol 1 pp 88–94 DOI 10.1109/APEIE.2014.7040794

[10] 1993 *Automatic tension control system for a 7-mill hot rolling mill 2500 of OJSC Magnitogorsk Iron and steel Works: Manual* (Davy McKee) p 19

[11] Shubin A G, Loginov B M, Khramshin V R, Evdokimov S A and Karandaev A S 2015 System of Automated Control of Hydraulic Screw-down Mechanisms of Plate Mill Stand *Proc of 2015 Int Conf on Mechanical Engineering, Automation and Control Systems (MEACS)* p 6 DOI 10.1109/MEACS.2015.7414858

[12] Stefanovich V L 1975 *Automation of continuous and semi-continuous broadstrip hot rolling mills* (Moscow: Metallurgy) p 208

[13] Tselikov A I 1962 *Theory of calculating the roll forces in rolling mills* (Moscow: Metallurgizdat) p 494

[14] Radionov A A, Karandaev A S, Evdokimov A S, Andryushin I Yu, Gostev A N and Shubin A G 2015 Mathematical modelling of the interrelated electric and mechanical systems of continuous sub-group of the rolling mill stands. Part 1 Development of the mathematical model *Bulletin of the South Ural State University. Ser. Power Engineering* vol 15 **1** pp 59–73

[15] Tsekikov A I, Tomleev A D and Zuzin V I 1982 *Theory of rolling: manual* (Moscow: Metallurgy) p 335

[16] Khramshin V R, Khramshina E A, Karandaev A S, Gasiyarov V R and Voronin S S 2017 Control Methods and Systems Providing Reduced Consumption Index at Rolled Product Manufacture at Plate Mill *Proc of the IEEE NW Russia Young Researchers in Electrical and Electronic Engineering Conf (EIConRus)* pp 1540–4 DOI 10.1109/EIConRus.2017.7910865

[17] Karandaev A S, Khramshin V R, Andryushin I Yu, Petryakov S A and Khramshin R R 2013 Method for correction of gauge interference of the head-strip section in a system for automated controlling of the thickness of a broad-strip hot-rolling mill *Russian Electrical Engineering* **84(8)** pp 441–5

[18] Shilyaev P V, Andryushin I Yu, Golovin V V, Radionov A A, Karandaev A S and Khramshin V R 2013 Algorithms of a digital automatic system for tension and loop control in a wide-strip hot-rolling mill *Russian Electrical Engineering* **84(10)** pp 533–41 DOI 10.3103/S106837121310009X

[19] Khramshin V R, Salganik V M, Zhilina V A and Yachikov I M 2015 Mathematical Model of Electromechanical Systems of Wide-Strip Hot-Rolling Mill Continuous Train *Proc of 2015 Int Conf on Mechanical Engineering, Automation and Control Systems (MEACS)* p 7 DOI 10.1109/MEACS.2015.7414892

Acknowledgments
South Ural State University acknowledges financial support from Ministry of Education and Science of the Russian Federation (grant No 13.9656.2017/8.9)

IOP Conf. Series: Materials Science and Engineering **361** (2018) 012021 doi:10.1088/1757-899X/361/1/012021

The Influence of Load Model on Transient Stability Limit Removal Time of Power System

Min Zhang, Zhaowen Luan, Xiangjun Zhao

School of Electrical Engineering, Shandong University Jinan250061,Shandong province,China

zhang881983@163.com

Abstract. In this paper, ieee10 machine 39 node system as the research object. Load model considering the composition ratio of the induction motor and changes the specific gravity impact on the power system transient stability limit voltage clearing time changes. The results show that the above parameters have an impact on the transient voltage stability of the power system. Considering the change of the composition of the load model can more fully analyze the transient stability. Therefore, the choice of the load model should be careful to avoid adverse consequences such as inaccurate evaluation due to inaccurate load model establishment.

1. Introduction

Before the grid is put into operation, the planning and testing after the input and control are the prerequisites for the safe operation of the grid. Planning, testing and control all need to establish a reasonable model for the load side. In recent years, many cases of transient voltage instability caused by inappropriate selection of load models have occurred in various parts of the world [1-6].This shows that the selection of the load model will have a more serious impact on the stability and calculation of the power system. Induction motor load is the main component of power system load and also the most important dynamic load, accounting for more than 90% of the industrial load, which is closely related to the transient voltage stability problem [7,8]. Document [9] proposes the concept of transient voltage stability limit removal time for an infinite bus system that contains an induction motor load. Document [10] further studies.Document [11] points out the importance of induction motor modeling in voltage stability. The importance of the load model in analyzing a series of dynamic behaviors of the power system has drawn more and more attention.

Based on the above reasons, this paper studies the effect of the polynomial integrated load model and induction motor model on the transient stability of power system.This article sets the type of circuit fault for the three-phase short circuit, the fault occurred at the beginning of the line.By simulating three different faults, the influence of the different combinations of the polynomial load model and the proportion of the induction motor model in the load model on the transient stability limit removal time of the power system was tested.

2. The load mathematical model

2.1. The polynomial load model

Content from this work may be used under the terms of the Creative Commons Attribution 3.0 licence. Any further distribution of this work must maintain attribution to the author(s) and the title of the work, journal citation and DOI.
Published under licence by IOP Publishing Ltd

The static load model of the relation between power and voltage amplitude are represented by polynomial function model. The polynomial model mainly reflects the relationship between voltage and frequency.Because it is composed of constant impedance Z, constant power P , constant current I three kinds of model and linear combination model ,It is also called the ZIP model.

2.2. The induction motor load model .

The induction motor model can be generally divided into three categories, one is the five order electromagnetic transient model and rotor electromagnetic transient, if you ignore the transient stator obtained is three order electromechanical transient model, and only consider the first-order mechanical transient model of mechanical transient.

3. The example analysis

3.1. Simulation System

The IEEE10 machine 39 node system structure shown in Figure 1.

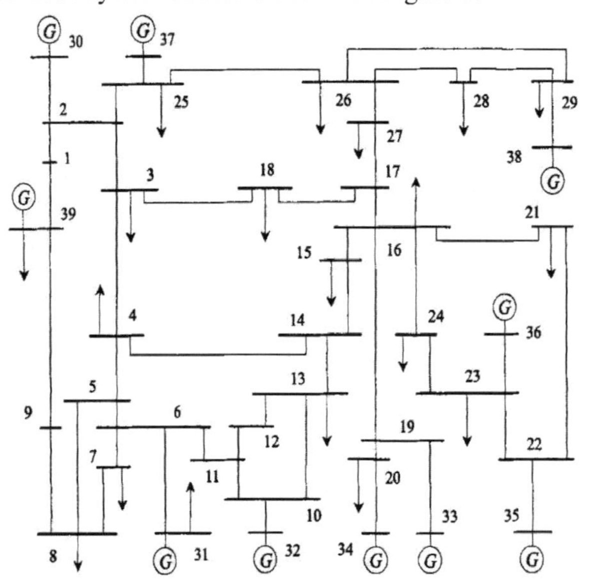

Figure 1. The IEEE 39 node 10 machine system structure.

This article considers the type of fault for three-phase short circuit, the fault occurred at the beginning of the line.By simulating three different faults(3 ~ 4 fault, 13 ~ 14 fault and 23 ~ 24 fault), the influence of the different combinations of the polynomial load model and the proportion of the induction motor model in the load model on the transient stability limit removal time of the power system was tested.

3.2. The result of changing the ratio of polynomial integrated load model .

In the case of three different fault cases, we respectively change the ratio combination of the comprehensive load polynomial model to get the corresponding fault limit removal time. The specific results are shown in Table 1, Table 2 and Table 3.

Table 1. The transient stability limit removal time when the location of the fault point for the branch 3 ~ 4.

The percentage of constant	The percentage of constant current （%）	The percentage of constant power （%）	Transient stability limit removal time （s）

impedance（%）			
100	0	0	1.2641
0	100	0	1.235
0	0	100	1.185
80	20	0	1.2641
80	10	10	1.2641
70	30	0	1.2641
0	4	96	1.185
0	5	95	1.195

Table 2. The transient stability limit removal time when the location of the fault point for the branch13 ~ 14.

The percentage of constant impedance（%）	The percentage of constant current（%）	The percentage of constant power（%）	Transient stability limit removal time（s）
100	0	0	1.2741
0	100	0	1.3040
0	0	100	1.2340
10	90	0	1.3040
12	88	0	1.3040
20	80	0	1.285
0	10	90	1.2541

Table 3. The transient stability limit removal time when the location of the fault point for the branch23 ~ 24.

The percentage of constant impedance（%）	The percentage of constant current（%）	The percentage of constant power（%）	Transient stability limit removal time（s）
100	0	0	1.235
0	100	0	1.235
0	0	100	1.175
90	10	0	1.235
80	20	0	1.235
70	30	0	1.235
0	2	98	1.175
0	4	96	1.185

Through the data in the table we can observe some phenomena and laws.

(1) Under the three kinds of fault conditions, the fault limit cut-off time corresponding to pure constant power load model is the smallest.

(2) There are two cases (3 ~ 4 and 23 ~ 24 failures), the maximum cutting time is the limit when the load model is constant impedance.

It can be seen that when the proportion of constant impedance reaches a certain value (in this case, 70%), the constant impedance model plays a leading role in the fault cut-off time. The proportional change of constant current and constant power basically has no effect on the fault cut-off time.When the ratio of the constant impedance is greater than this value, the fault limit resection time is no longer sensitive to the change of the ratio of the constant impedance. In this case, when the ratio of the constant impedance is greater than 70%, the fault limit resection time is the largest and remains unchanged.However, when the proportion of constant impedance is very small, the model of constant current and constant power plays a leading role in the fault removal time, and the fault removal time is very sensitive to the ratio of constant current and constant power. One percent change will cause the fault cut-off time to change.

(3) When the 13 ~ 14 branch failures, the constant current load model fault removal time limit the maximum.Similarly, the percentage change of constant current has a greater impact on the fault cut-off time.

3.3. The result of changing the induction motor ratio in the load model.

In the case of three different fault cases, the proportion of induction motor load models were changed respectively, and the corresponding results of the fault limit resection are shown in Table 4.

Table 4. The fault limit cut-off time when the proportion of induction motors in the load model is different(The generator considers the damping mode with no damping and constant potential respectively).

Induction motor load ratio	Fault line	3～4	13～14	23～24
25%	No damping	0．334	0．174	0.214
	Damping	0．354	0．214	0.224
50%	No damping	0．304	0．104	0.154
	Damping	0．324	0．144	0.164
75%	No damping	0．184	0．074	0.094
	Damping	0．194	0．114	0.104
100%	No damping	0．054	0．034	0.074
	Damping	0．074	0．074	0.894

We can see from the above table the following conclusions.

(1) When the same fault and the proportion of induction motor model are the same, the fault limit removal time without damping must be smaller than when there is damping.

This shows that the impact of generator with or without damping and induction motor load model has nothing to do.

(2) Whether the generator considers the damping or not,as the proportion of the induction motor model increases, the fault cut-off time decreases.

This is because the power system transient stability calculation is the analysis of electromechanical transient process.It mainly reflects the unbalanced relationship between the mechanical power and the electromagnetic power of rotating electrical components in the system.Induction motor load model describes the motor rotor equation of motion, which directly reflects the imbalance between mechanical power and electromagnetic power.The greater the proportion of induction motors in the load model, the greater the component of the relationship between mechanical power and electromagnetic power, so the greater the effect on the fault cut-off time.

4. Conclusion

The proportional combination of the polynomial integrated load model and the proportion of the induction motor model have different effects on the power system fault removal time.Using different load models will directly affect the results of the transient stability calculation, thus affecting the power system planning, design, operation and other decisions based on this.

The contents of this paper have some reference value for further research on load dynamic modeling and transient voltage stability.

Because this paper mainly analyzes the influence of power system load model on transient voltage stability,the paper does not consider the impact of certain factors such as generator excitation of the head-end power system.

5. References

[1] DY H 2004 *J.Power System Technology* **28 (21)** 1-5
[2] Hu X 2003 *J.Power System Technology* **27(9)** T2-T6
[3] Yin Y，Guo J,Zhao J *J.Power System Technology* 2003 **27(10)**8-11
[4] Tang B 2003 *J.Power System Technology* **27(11)**1-6
[5] Han Z，Cao Ya 2004 *J.Power system security to ensure reliable delivery* **28(9)**1-6
[6] Qu J，Guo J 2004 *J.Power System Technology* **28(21)**60-62
[7] Taylor C W 1994 *M.Power system voltage stability* **New York** Mc Graw-Hill
[8] Taylor, C. W. 1993 *J.American Journal of Neuroradiology* **25(25)** 1862-3
[9] El-Sadek, M. Z. 1991 *J. Electric Power Systems Research* **21(1)**9-16
[10] Duan, X. D. 1995 *M.Journal of Huazhong University Ofence & Technology*
[11] Mechenbier, J., Ellis, A., Curtner, R., & Ranade *S 2005 C.Power Engineering Society General Meeting* **IEEE 2005** 1611-1618 Vol.2

IOP Conf. Series: Materials Science and Engineering **361** (2018) 012022 doi:10.1088/1757-899X/361/1/012022

Research and analysis of dynamic equivalence of power system

X J Zhao, Z W Luan and M Zhang

School of Electrical Engineering, Shandong University, Jinan, Shandong Province, China

sukeez@qq.com

Abstract. In recent years, with the expansion of power grids, power system interconnection has also expanded. However, it brings many problems to the operation and management of power systems. The dynamic equivalence of power system is a kind of equivalent method of power system, which can keep the area of power system that needs to be studied unchanged and simplify the external area equivalently. Dynamic equivalence is divided into the cohomology equivalence method, the model equivalence method and the equivalent method of estimation. The dynamic equivalence of large-scale power system can improve the calculation accuracy, reduce the computational complexity and greatly improve the grid efficiency. This paper summarizes the dynamic equivalent methods of power systems in recent years, introduces the ideas and methods of various methods, and carries out the simulation analysis.

1. Introduction
Study on the equivalent method of power system, have a very important practical significance on the off-line analysis and on-line analysis. With the development of power grid construction, the formation of interconnected power system brings many new problems to the calculation of power system planning and operation mode. It is very difficult to get all the accurate real-time information of the whole external system when the analysis and computation of the interconnected system are in various running states, but the scale of the mathematical model of the system must match with the real-time information obtained. Therefore, it is necessary to deal with some non-observable parts of the external system as external equivalents.

In the process of large-scale system analysis, it is generally only interested in one of the local systems, called the research system, also known as the internal system. For the area far from the research system, it can be appropriate to simplify the treatment to save manpower and material resources, as long as the dynamic impact of the study system to maintain the same, its internal structure is not necessary to be investigated in detail. This area can be called an external equivalent system. Systematic simplification highlights the main contradiction, which is necessary to master the main features of the research system[1-3].

2. Several methods of dynamic equivalence
The method of dynamic equivalence is closely related to the actual physical problems which are studied after equivalence. Generally, the problems can be divided into three types[4-6].

Content from this work may be used under the terms of the Creative Commons Attribution 3.0 licence. Any further distribution of this work must maintain attribution to the author(s) and the title of the work, journal citation and DOI.
Published under licence by IOP Publishing Ltd

IOP Conf. Series: Materials Science and Engineering **361** (2018) 012022 doi:10.1088/1757-899X/361/1/012022

- Transient Stability Analysis of Large Power System. Its feature is that the system structure and parameters are known, the system transient process should be analyzed in the case of large disturbances, the system was strong nonlinear. The requirement for dynamic equivalence is that the research system should have a similar rocking curve before and after the equivalence.
- Dynamic stability analysis of off-line large power system. Its feature is that the system structure and parameters are known, the system transient process should be analyzed in the case of small disturbances. The requirement for dynamic equivalent is that the research system should have the same or similar motion pattern distribution before and after the equivalence.
- Apply the measured data to the system equivalence. Its feature is that system conditions and structure are various, but there are a lot of measured values available. It is necessary to identify the equivalence of the external system quickly so as to provide sufficient external information support for the on-line stability analysis of the system, and ensure that the research system has similar dynamic safety analysis results before and after the equivalence.

Accordingly, equivalence methods can be divided into three categories:

- Coherency-based equivalents based on relevance concept. According to the transient stability analysis of large power systems, those generators with the same frequency and similar frequency swing (i.e., the same tumbling) are grouped together. All generators in each group are aggregated into one or two equivalent machines. Through the network simplification, a large number of load nodes can be eliminated. At present, it is limited to the combination of constant power, constant impedance and constant current load on the basis of reserved power. Coherency-based equivalents have the advantages of large physical transparency and equivalent system components. The model is the actual power system component model, which can directly apply the original transient stability analysis software. It can adapt to the system of strong nonlinear and large perturbation analysis, and can be applied to large-scale system equivalent; the equivalent value can be carried out on the minicomputer, the speed is very fast; dynamic equivalent precision control is more convenient, the practical application shows that the application effect is better. But coherency-based equivalents also has some shortcomings: Firstly, homology group division and disturbance of the location, type and other factors; Secondly, the network simplification and elimination of the phase shifter operation will bring some errors to the dynamic equivalent; In addition, the coherent generator polymerization is more complex, will lead to a certain polymerization error. In recent years, there have been new developments in the identification of homology groups.
- Modal equivalents based on eigenvalues. It is the use of compression system motion mode number of ways to reduce the physical order of the system, thereby reducing the demand for computing. This method is not only the design of the system, the integration is very effective in the system stability simulation and dynamic safety analysis is also quite valuable. The advantage of modal equivalents is that once the dynamic equivalence of the external system is obtained by this method, the same equivalent result can be used to analyze the large number of accidents in the research system, as long as the fault point is not close to the boundary point. However, due to the mechanism of modal selection is not very clear, it is only limited to obtain the external system state equation model decomposition, try to improve the simulation speed.
- Online recognition based on measured values. Both of these methods have large computational complexity and need to know the difficulty of all the data of the external equivalent system, and often have the small disturbance linearization of the system, and it is difficult to meet the engineering requirements in time and precision. In order to overcome these difficulties, it is considered to use the boundary line to measure the information through the identification method to obtain the external region of the equivalent model. Unlike modal equivalents and coherency-based equivalents, online identification does not require detailed data from external systems, which is what people expect. However, due to the nonlinear time-

varying characteristics of power systems, as well as the special requirements of online identification, the realization of this method there are some difficulties.

3. Equivalent Model

In the early study, the structure of the two regions is generally used, and the equivalent model is a single load. As the model is simple, coupled with the constraints of simulation technology, the accuracy of numerical calculation cannot meet the needs of engineering practice. In recent years, on the basis of this gradually carried out on the multi-regional interconnection system equivalent problem, and put forward a variety of different equivalent structure model[7-10], respectively, as follows.

The actual contact line model retains the actual tie-line, which equates each regional system into a single-load model, as shown in figure 1.

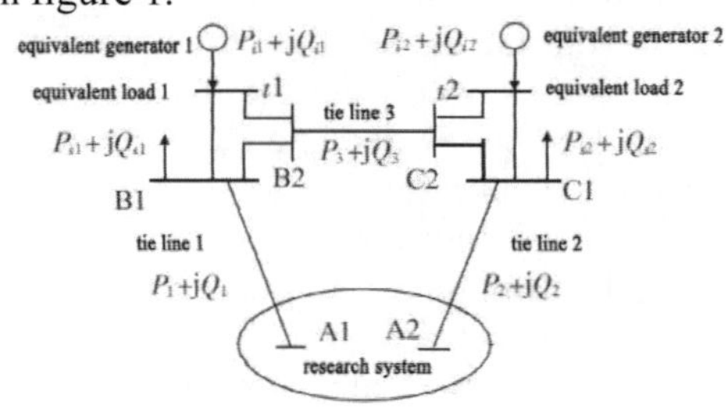

Figure 1. Actual contact line model.

The equivalent contact line model is obtained from the equivalent model in figure 1 for further processing, as shown in figure 2. This can be further simplified for the equivalent system, of course, compared with the system in figure 1, the accuracy will be reduced.

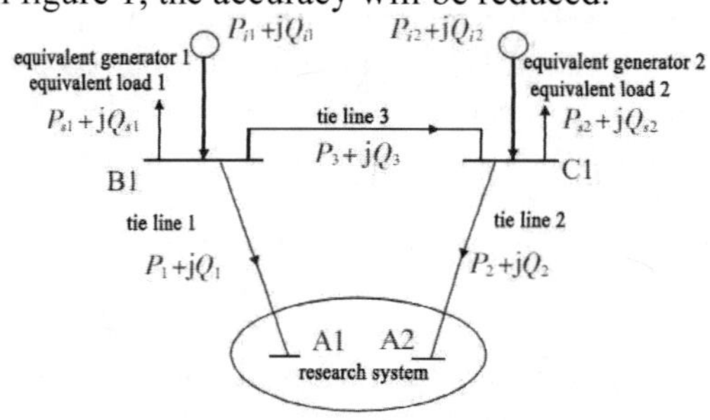

Figure 2. Equivalent contact line model.

For large-scale equivalent systems, sometimes single-load model is difficult to meet the accuracy requirements of the analysis. In addition, taking into account the external equivalence of the system and the distance between the research system is not exactly the same, if necessary, the equivalent model can make appropriate adjustments to divide the buffer, as shown in figure 3. Buffers can improve the accuracy of a certain extent, but after the division of the buffer, you need to do more analysis of external systems.

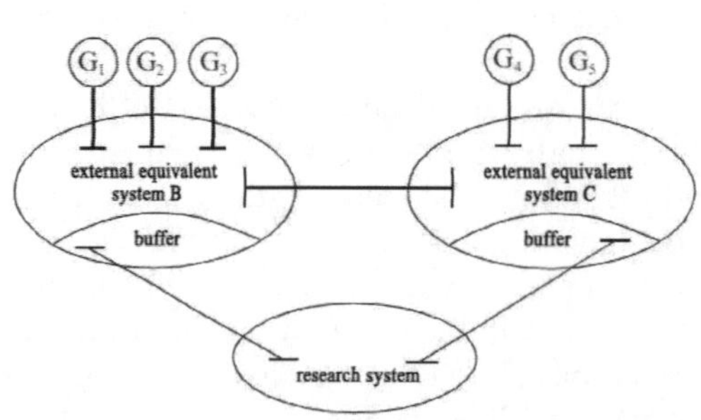

Figure 3. Multi-machine equivalent model with a buffer.

IOP Conf. Series: Materials Science and Engineering **361** (2018) 012022 doi:10.1088/1757-899X/361/1/012022

The results show that the actual contact line model is more suitable for the theoretical equivalent method, and the equivalent contact line model is more suitable for identifying the equivalent method. The multi-machine equivalent model with a buffer can be applied to the problem analysis which needs more precision.

4. Simulation results and discussions

Figure 4 shows the IEEE39-node power system single-line diagram. The system has 10 machines, 19 loads, and node 31 is the balance node. In the PSASP transient stability calculation, all ten generators adopt a third-order model. The system nodes 2, 17 and 18 are the boundary of the equivalence network, and the nodes 1, 3, and 16 are the boundary of the research network. In this part, the power flow calculation and the transient stability calculation of the equivalent network are compared and analyzed respectively.

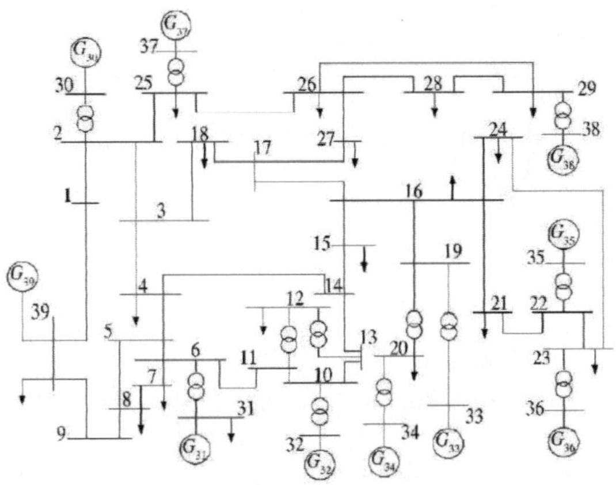

Figure 4. IEEE39 bus power system

By the equivalent calculation, the constant potential of the equivalent machine is 1.598, the value of the equivalent potential angle is 43.12 ° and the inertia constant is 17.9s. The power flow is calculated using the equivalent data set and compared with the equivalent network trend, as shown in Tab. 1. In Tab. 1, the transmission of active and reactive power is the power of the equivalent network boundary to the research network. The positive value indicates that the equivalent network transmits the power to the research network. The negative value indicates the equivalent network to absorb power from the research network.

Table 1. Boundary flow of the IEEE39 bus system before and after equivalence

Boundary Node		Before Equivalence		After Equivalence	
Equivalence system	Research system	active power	reactive power	active power	reactive power
2	1	1.1598	-0.2689	1.1599	-0.2969
2	3	4.0679	0.9917	4.0701	0.9917
17	16	-1.2778	0.1091	-1.2769	0.1091
18	3	-0.2586	0.0192	-0.2586	0.0192

As can be seen from Table 1, the power of the equivalent front and back boundary nodes is almost unchanged, and the voltage phase angle of each node in the research network is similar. EPRI E 'equivalent calculation can meet the requirements of steady-state power flow calculation in engineering calculation, and can save the calculation time by the equivalent research network.

The transient analysis of the equivalent system is carried out by assuming a serious fault of the three-phase ground short circuit. Observe the relative angle δ of the rotor of the generator, and observe the near point of the distance equalization subsystem and the far point of the fault point distance equivalent subsystem.

Near the point selection bus 16 at the exit, two phase earth fault , the fault occurred at 0.2s, and cut off at 0.35s, observe the generator 35 and 39 rotor relative angle curve Far point selection bus bar 11, three phase earth fault , the other experimental conditions and near point of failure the same.

Figure 5 shows the change of the relative angle of the rotor before and after the near-point fault. Figure 6 shows the relative value of the rotor before and after the fault.

By contrast, it can be seen that the change of the relative angle of the rotor of the system after the original system and the simplified system is consistent within 2s after the system failure, and does not affect the transient stability analysis.

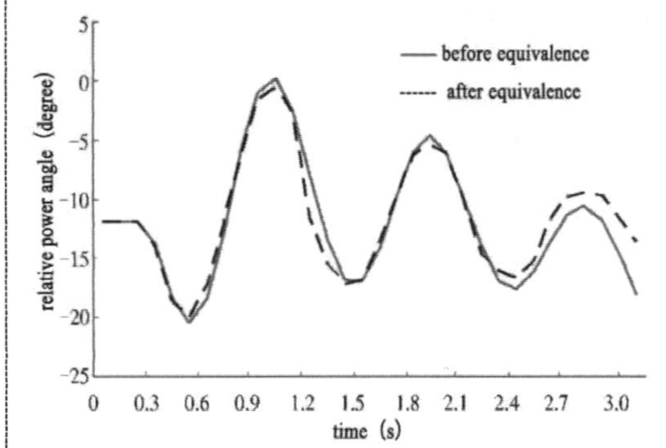

Figure 5. The curve of relative rotor angle before and after equivalence at vicinity

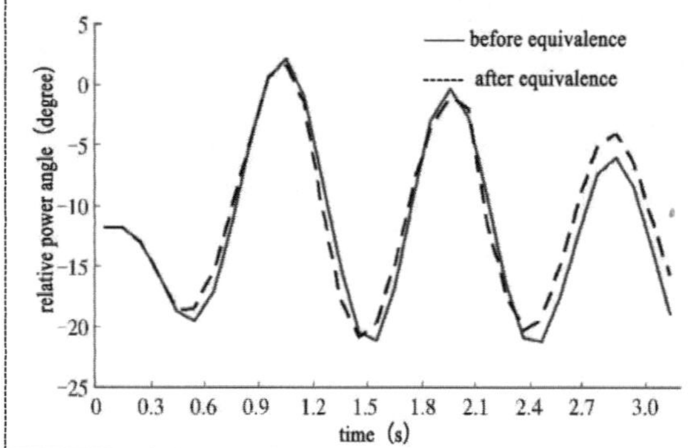

Figure 6. The curve of relative rotor angle before and after equivalence at distance

5. Conclusion

For simplified calculation, we often simplify some areas that do not need intensive study in power system analysis. By using the PSASP software to process dynamical equivalence calculation in typical example and actual large scale power networks, this paper analyzes the advantages and disadvantages of PSASP dynamical equivalence function. It is convenient to use. It could calculate multiple equivalence and the calculated results are suitable for transient analysis. But in some situations the EPRI equivalent cannot get results, and the equivalent calculation interrupts. The specific causes and improvement measures need to be further studied.

References

[1] Ni Y X 2002 The Dynamic Theory and Analysis of Power System (Beijing: Tsinghua University)

[2] Xu J B, Xue Y S,Zhang Q P and Wang D X 2005 *J. Automation of Electric Power System.* **29** 91–5

[3] Kim J Y and Won D J et al 2001 *Int. Conf. on Power Engineering Society Summer Meeting(Canada)* vol 2 pp 973-977

[4] Wang G and Zhang B M 2006 *J. Power System Technology.* **30** 21-6

[5] Li J, Chen H and Li D L 2007 *J. Guangdong Electric Power.* **20** 1-4

[6] Yang J P and Xu Z 2005 *J. Power System Technology.* **29** 68-71

[7] Ju P, Wang W H, Xie H J and Zhou H Q 2007 *J. Proceedings of the CSEE.* **27** 29-34

[8] Zhou Y H, Li X S, Hu X Y and Lin T B 1999 *J. Proceedings of the EPSA.* **11** 29-33

[9] Li X Q Shao Y H and Zhang Q 2008 *J. Electrical Engineering.* **2** 34-6

[10] Yang N G, Wang B H, He N Q, Wu F and Ju P 2001 *J. Automation of Electric Power Systems.* **25** 48-51

IOP Conf. Series: Materials Science and Engineering **361** (2018) 012023 doi:10.1088/1757-899X/361/1/012023

The Usage of Overall Equipment Effectiveness in Measurement to Improve Efficiency and Increase Productivity of Process for Packaging Cigarettes

Bunsong Khamon

Department of Industrial Engineering, Faculty of Engineering
Ramkhamhaeng University, Bangkok, 10240, Thailand

Email: bunsong4051@ru.ac.th

Abstract The purpose of this research is to improve efficiency and productivity of production process of cigarettes of factory 5 by select the packing process cigarette 2 to study, and use the Overall Equipment Effectiveness (OEE) in measurement to improve. The results of the study showed the main cause of Overall Equipment Effectiveness and productivity was declined including organizing and labor, machinery, raw materials. The proposed guidelines of improvement are as follows. (1) Organizing and Labor, including the Job duties and responsibilities must be clearly identified. Staff training to improve their knowledge and expertise must be provided. (2) Machinery, including implementation of Total Productive Maintenance (3) Raw materials, including discovering the appropriate conditions of cigarette packing method by the experiment of central composite design. After the application of the mentioned method, showed the Loading Efficiency Index value was higher to 92.59%, the Machine Efficiency Index value was up to 84.17% and the Quality Rate Index value was increased to 98.88% as a result, the Overall Equipment Effectiveness increased from 51.05% to 77.07%, production efficiency was increased from 52.27% to 79.66% and productivity was increased from 9,899 packs per hour to 15,058 packs per hour or was increased to 52.12%

1. Introduction

[1] In any business we have to produce with quantity, quality and cost set by production standards. The sample factory that is the cigarette industry is experiencing productivity quality problems and falling productivity. It's cause of increase of cost, due to overtime to increase the number of planned production. Sometime cigarettes out of stock make lack of customer trust. So that the factory have to continuous improvement and development follow corporate development strategy year 2017-2021. By the aim to improve efficiency and increase effectiveness, personnel development and organization. Improve product quality, develop new production. In order to meet the purpose of the sample factory as a guideline to follow the plan to improve efficiency and increase effectiveness, so that the research was conducted. Based on a retrospective study of fiscal year 2016, found that the amount of output at normal working hours the average was relatively low 61.02% compared with the planned production volume. Plan division was set the output target not less than 90%.

Content from this work may be used under the terms of the Creative Commons Attribution 3.0 licence. Any further distribution of this work must maintain attribution to the author(s) and the title of the work, journal citation and DOI.
Published under licence by IOP Publishing Ltd

IOP Conf. Series: Materials Science and Engineering **361** (2018) 012023 doi:10.1088/1757-899X/361/1/012023

This research has chosen the packing process cigarette 2 from factory number 5 to study because of manufacturing process was bottleneck, include low efficiency and low productivity to study to guide to improve the efficiency and productivity following the target.

2. Objective

To improve the efficiency and productivity process of packing cigarette 2 of factory number 5 by using the overall effectiveness value of the machine is a measure of improvement by setting production increase efficiency targets of not less than 75% when compared with the planned production amount.

3. Research Methodology

3.1 General information of the sample factory
The sample factory has 3 factories and produce brands of filter cigarettes is divided into 17 brands for the example Kor. 90 and Kor. American Style and Kor. Light. And produce non-filter cigarettes 5 brands such as Kor. 33, Sor. 33 and Ror. 33 etc. Based on the previous information of fiscal year 2012-2016 was found that production capacity is about 5,614.18 million roll per year. The production details are as following the factory no. 3 was produced about 3,099.59 million roll cigarettes per year accounted for 55.21%. Factory no. 4 was produced about 1,492.81 million roll cigarettes per year accounted for 26.59% and factory no. 5 was produced about 1,021.78 million roll cigarettes per year accounted for 18.20%. This research brings previous fiscal year 2016 data to study, it was found that the production of factory no. 5 was a low performance on the average 57.27% compared with the planned production amount. So that we choose factory no. 5 to study because of the minimum efficiency. which is a cigarette factory with modern machinery that the process of producing cigarettes is divided into two parts. Such as tobacco manufacturing process, roll and packing processes. Tobacco manufacturing process is the responsibility of the tobacco division. Start from the preparation of raw materials, the process of roll and packing cigarettes is the duty of the roll and packing division. Takes duty to receive tobacco from tobacco division by the suction pipe into the roll machine and processed tobacco into cigarettes.

3.2 Analyze the problem of the process
Based on the study of production conditions of factory 5 have two processes of cigarette production. It was found that the process of producing cigarettes is a bottleneck from the process of packing the cigarette into the package. From the data collection between October and November 2016, it was found that the process of packaging cigarettes 2 lowest performance. The average equal 1.96 million packs per month compared with the planned production was 52.27% and the average was 9,899 packs per hour. Which is the efficiency and productivity are low this process uses the cigarette packaging process 2 as a sample process. To improve and guide to increase the effectiveness of cigarette production.

3.3 Analysis of the causes of decline productivity
From the study process of packing a cigarette 2, it has been found that the main reason of the efficiency and productivity falling down. The cause of organization and labor, the cause machine and raw material reasons.

3.3.1 Analyzing the causes of organization and labor from the study of the organizational structure of the roll and packing department. Found that not used the organizational chart officially, made lack of clear flow of work order, lack of follow-up control and evaluation. **[2]** Which is the principle of corporate management. Summarize causes of current organization and labor as a guideline for improvement as the following

1) Lack of communication and coordination between agencies.
2) Lack of used formal organizational charting or written assignments to assign supervisory lines and description of each position.
3) Staff lack training support to make their work more efficient.

3.3.2 Analysis cause of machine manufacture of cigarettes, the department of roll, packing and boxes packing are the main machine. Due to the manufacturing process, the cigarette produces was bottleneck at the packing with packing machine. From the study have collected the information about the loss of cigarette packing machine in terms of time, performance and quality. This will be used as a performance index of indicator loss as Loading Efficiency, Machine Efficiency, Quality Rate and Overall Equipment Effectiveness. **[3]** definition of OEE that Total Productive Maintenance: TPM is OEE it include operations, maintenance, and management of tools and Overall Equipment Effectiveness (OEE) it is a measure of efficacy and effectiveness of machine operation by the equation is calculated as follows: OEE = Loading Efficiency (%) × Machine Efficiency (%) × Quality Rate (%)
3.3.3 Raw material analysis, raw materials used to produce cigarettes of roll and packing department to become cigarette are divided into two categories as tobacco and wrappers, tobacco are ingredients that are cut into striated and through the process of preparation of the tobacco division. The wrapped material are rolling paper, filter paper, Rolling seam adhesive paper wrapper, glue stick etc. Sometimes, when the glue or some materials such as seam pack butts, filter paper, but not used properly. That the filter removes from the cigarette. The pack seam breaks apart. This is the main causes a loss of time and quality. Due to the adhesive material parameters seam the pack and improper parameter setting of the cigarette packaging machine. From data collected since October to November 2016 showed that on average, the proportion loss is 0.0597. Limited target to reduce of broken seam pack 70% by starting from the assessment of the stability of the process packing cigarette. Which is in control or not, and **[4]** configured to test data abnormality on the control chart by reference the United States Automotive Industry Data Abnormality Test (AIAG). Based on the analysis of proportional control charts. The loss contains all the data that plot on the control chart in the line control and the information is random. It can be concluded that the proportion of loss and stability of the process is stable and the data is random. So the problem was analyze and modified.

3.4 Improving efficacy production process

3.4.1 Organizational and Labor Improvement. In implementing the improvement was taken organizational structure of roll and packing department is used officially or written in writing. Then doing a job description of each job is provided let each person know the role of the job by identifying position name, the duties and responsibilities of each position clearly and training staff to increase their expertise.
3.4.2 Machine improvement
Roll and packing department in the manufacture of cigarettes are mainly used machines which high cost imported from abroad its lifetime over 10 years. For this reason, it is important to keep maintenance machinery as specially. Especially in the process of roll cigarette to packing as the bottleneck of the production process Therefore, **[2]** take maintenance techniques by used TPM for all participant to apply to improve. In order to Applying this technique. Starts with the management to support and accept the technique first. After that sub-group activities will be established in the production process. Then embark on the following 4 steps as the following

1) Determine the area and set up a team to improve.
2) Basic knowledge training required for the concerned person
3) Gather data and issues before improve then present to team members, develop team and maintenance teams to know by yourself to understand and solve the problem of the production process.

4) Apply TPM techniques to solve problems of production lines after that, each group presents the results which is applied to the management team.

3.4.3 Raw material improvement

Based on the raw material analysis it was found that the raw materials used packing had problems in quality and standard of parameters material of adhesive seam of pack. Including to parameter adjustments the appropriate of the packing machine. Because the parameters are very effective for process of packing a cigarette. From the problem of parameters standard, this research will study the appropriate parameters. The process of packing a cigarette by experiment design the steps are as following.

3.4.4 Analysis of characteristics and causes of loss

From the data collection Between October and November 2016, the average of productivity was 1.96 million packs per month. The amount of loss is 0.117 million packs per monthly. Accounted for the proportion of loss ratio of 0.0597 or 5.97% They can be classified by loss type as following (1) The pack seams was broken 97,660 packs cigarettes, accounting for 0.8347 (2) Cigarettes was broken and no filter bottoms, 19,040 packs, accounted for 0.1627 of all loss. Based on the information, there are a lot of loss in broken seam pack, **[5]** therefore we used designed 2^k Factorial and Central Composite Design to applied in order to decrease the loss of broken seam package 70%. By starting to find the cause of package seam broken from brainstorming of people involved in the process of producing cigarettes. Using the cause and effect diagram, to know the cause of the problem is identified. Summarized as following

1) Moisture of tobacco at present the tobacco room sets the standard moisture content of cigarettes at 12%
2) Current adhesive material used the standard is to set the glue quantity at 40% stickiness 1800 - 3500 Cps at 27 ± 2 °C.
3) The current speed of the packing machine is 300 pack per minute
4) The temperature of the seam packing currently sets the standard temperature at 200 °C.
5) The pressure seam adhesive of packing at the present set at 5 psi and wrapper type C.

The researcher and the improvement team brainstormed and considered together and prioritized the cause of the loss by using process analysis techniques, **[6]** the form of loss and process effects (Process FMEA). The result of the analysis that conclude the number of risk levels (RPN) are in descending order. Such as the temperature of the seam packing is 648. The adhesive material (% adhesive) is 504. Pressure adhesive seams have a value of 343, the humidity of the tobacco is 86, and the wrapping paper has a value of 75, respectively. The speed of the cigarette packing machine cannot be considered due to the productivity. This research has introduced the factors that affect the loss that have a higher risk than the 100 value to study such as the temperature of the seam packing machine for packing seam, types of glue, pressure adhesive, Experimental control variables is moisture of tobacco is maintained at 12%. The speed of the packing machine remains at 300 pack per minute and wrapped paper still use as the type C. Because the researcher and the improvement team found that it was not the effective of loss the experiment was set up and controlled as described above.

3.4.5. Determination of suitable conditions by experimental design.

Experimental design by use two factorial model with 2 levels (2^k Factorial) This step is a preliminary experiment with two factorial design techniques with 2 level that have increasing center in order to determine these three factors affect with loss in any form. After screening Factor with FMEA technique from the researcher and the team has improved, that should be define the following factors: Factor A is the temperature seam pack with 2 level: Low (180°C) and High (220°C). Factor B is the pressure seams was set with 2 levels: Low (4 psi) and High (6 psi). Factor C is adhesive material was set a two-level: Low (glue 30%) and High (glue 50%). Researchers and team provided factor for experimentation and control of variables in the experiment response variable, **[7]** Proportion of loss of each run will be randomly measure of sample size as $n=(Z_{1-\alpha/2}+Z_{1-\beta})^2/N\delta^2$ by Power = 0.90 and Alpha = 0.05, Calculation Units per run = 354 packs, To avoid too many zeros to user Units

IOP Conf. Series: Materials Science and Engineering **361** (2018) 012023 doi:10.1088/1757-899X/361/1/012023

per run = 400 packs then bring the value to be converted to the defect proportion. The results of 2 level factorial experiments as shown in Table 1.

Table 1. Session analysis Factorial Fit: defect proportion versus A (temperature), B (pressure), C (adhesive). Estimated Effects and Coefficients for defect proportion (coded units)

Term	Effect	Coef	SE Coef	T	P
Constant		0.05304	0.000082	649.57	0.000
A (temperature seam)	-0.02792	-0.01396	0.000082	-171.01	0.000
B (pressure)	-0.00522	-0.00261	0.000082	-32.00	0.001
C (adhesive)	-0.01243	-0.00621	0.000082	-76.09	0.000
A (temperature seam) * B (pressure)	-0.00893	-0.00446	0.000082	-54.65	0.000
A (temperature seam) * C (adhesive)	0.00508	0.00254	0.000082	31.08	0.001
B (pressure) * C (adhesive)	-0.00162	-0.00081	0.000082	-9.95	0.010
A (temp.) * B (pressure) * C (adhesive)	0.00357	0.00179	0.000082	21.89	0.002
Ct Pt		-0.00207	0.000156	-13.25	0.006
Curvature					0.006
S = 0.000230940 R-Sq = 100.00% R-Sq(adj) = 99.98%					

Table 1: Results of the Experimental Analysis on Session to find out what factors influence the process which use the p-value from the Estimated Effects to choose which factors have a significant influence This research determined the significance level of 0.05 The three main factors were Factor A (temperature seam), Factor B (pressure) and Factor C (adhesive). As well as the joint influence between Factor A(temp.seam)*B(pressure), A(temp.seam)*C(adhesive), B(pressure)*C(adhesive), A(temp.)*B(pressure)*C(adhesive). Significant statistical influence. Because the p-value is less than 0.05.

3.4.5.1. Central Composite Design: CCD

Table 1: Center Point (Ct Pt), Curvature, P-value less than 0.05 show that the variables studied have linear curvature relationships. **[8]** Therefore use the design of the CCD experiment this is an experimental for studying variables with linear curvilinear relationships but the number of Runs is not much, and additional experiments from 2^k Factorial. Therefore, the main factors affecting the three response variables are temperature seam, adhesive, pressure to analyze and design experiment. In order to try to find the nearest point with to the best production by additional experiments were conducted using the base of the 2^k Factorial. To find the optimal factor level that minimizes the amount of waste that occurs in the process and set α at 0.05. Which can display level of factor, experimental results, Analysis of Variance for defect proportion, and Show relationships that affect the proportion of loss (Estimated Regression Coefficients for defect proportion using data in uncoded units) as shown in Tables 2, 3, 4 and 5 respectively.

Table 2. Factor level of Central Composite Design

Factor	Symbol	Factor Level					Unit
		-α	-1	0	+1	+α	
temperature seam	A	166	180	200	220	234	degree Celsius
pressure	B	3.3	4	5	6	6.7	psi
adhesive	C	23	30	40	50	57	Percent

Table 3. Experiment result (CCD)

Factor A	Factor B	Factor C	defect proportion
temperature seam	pressure	adhesive	
180	4	30	0.0713
220	4	30	0.0508
180	6	30	0.0802

220	6	30	0.0347
180	4	50	0.0590
220	4	50	0.0415
180	6	50	0.0575
220	6	50	0.0293
200	5	40	0.0511
200	5	40	0.0507
200	5	40	0.0511
166	5	40	0.0727
234	5	40	0.0304
200	3.3	40	0.0500
200	6.7	40	0.0453
200	5	23	0.0682
200	5	57	0.0484
200	5	40	0.0479
200	5	40	0.0536
200	5	40	0.0508

Table 4. Analysis of Variance for defects proportion

Source	DF	Seq SS	Adj SS	Adj MS	F	P
Blocks	1	0.000002	0.000001	0.000001	0.10	0.754
Regression	9	0.003361	0.003361	0.000373	62.72	0.000
Linear	3	0.003013	0.003013	0.001004	168.69	0.000
A (temperature seam)	1	0.002448	0.002448	0.002448	411.15	0.000
B (pressure)	1	0.000061	0.000061	0.000061	10.20	0.011
C (adhesive)	1	0.000504	0.000504	0.000504	84.72	0.000
Square	3	0.000131	0.000131	0.000044	7.36	0.009
A (temp_seam) * A (temp_seam)	1	0.000000	0.000001	0.000001	0.25	0.632
B (pressure) * B (pressure)	1	0.000026	0.000016	0.000016	2.70	0.135
C (adhesive) * C (adhesive)	1	0.000105	0.000105	0.000105	17.64	0.002
Interaction	3	0.000216	0.000216	0.000072	12.10	0.002
A (temp_seam) * B (pressure)	1	0.000159	0.000159	0.000159	26.76	0.001
A (temp_seam) * C (adhesive)	1	0.000052	0.000052	0.000052	8.65	0.016
B (pressure) * C (adhesive)	1	0.000005	0.000005	0.000005	0.89	0.371
Residual Error	9	0.000054	0.000054	0.000006		
Lack-of-Fit	5	0.000037	0.000037	0.000007	1.82	0.291
Pure Error	4	0.000016	0.000016	0.000004		
Total	19	0.003416				
S = 0.00244004	R-Sq = 98.43%	R-Sq(adj) = 96.69%				

Table 5. Estimated Regression Coefficients for defects proportion using data in uncoded units

Term	Coef
Constant	0.1305060
Block	0.0001782
A (temperature seam)	-0.0003809
B (pressure)	0.0563581
C (adhesive)	-0.0049039
A (temperature seam) * A (temperature seam)	0.0000008
B (pressure) * B (pressure)	-0.0010592
C (adhesive) * C (adhesive)	0.0000271
A (temperature seam) * B (pressure)	-0.0002231
A (temperature seam) * C (adhesive)	0.0000127
B (pressure) * C (adhesive)	-0.0000813

3.4.5.2 Summary of Central Composite Design

From Table 4, the results of Central Composite Design it was found that the temperature seam, pressure, adhesive, (adhesive)*(adhesive), (temperature seam)*(pressure) and (temperature seam)*(adhesive) the p-value is less than 0.05. It can be concluded that these factors affect the amount of loss in the cigarette packing process. And the P-value of Lack-of-Fit is 0.291 this is more than the significance value level 0.05. Means a regression model for the loss proportion is appropriate Significantly, by R-Sq (adj) = 96.69%.Thus, a regression model was developed for the waste proportion as equation (1) and check the statistical hypothesis by graph Residual Plots that not found conflict with the statistical assumptions.Table 5 shows the relationships that affect the proportion of loss and it can be modeled regression equation that shows the relationship between various factors. Which influence on the proportion of loss. As equation using data in uncoded units (1)

$$Y_{(minimum)} = 0.130506 - 0.0003809(temp\_seam) + 0.0563581(pressure) - 0.0049039(adhesive) - 0.0002231(temp\_seam)*(pressure) + 0.0000127(temp\_seam*(adhesive) + 0.0000271(adhesive)*(adhesive) \quad (1)$$

3.4.5.3 Determining the optimal factor value from the experiment

Based on the equation (1), find the proportion value of loss generated by the experiment analysis with the Response Optimizer will be get the best appropriate values for all three factors are shown in Figure 1.

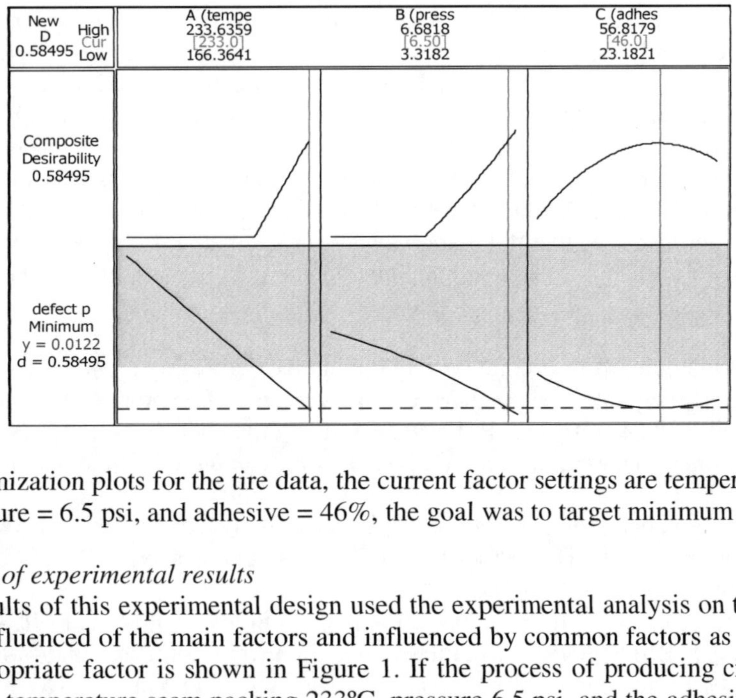

Figure 1 Optimization plots for the tire data, the current factor settings are temperature = 233 °C, pressure = 6.5 psi, and adhesive = 46%, the goal was to target minimum at 0.0122.

3.4.5.4. Summary of experimental results

The conclude results of this experimental design used the experimental analysis on the Session. Helps to translate the Influenced of the main factors and influenced by common factors as shown in Table 4, and find the appropriate factor is shown in Figure 1. If the process of producing cigarette 2 step the packing using the temperature seam packing 233°C, pressure 6.5 psi, and the adhesive material has an adhesive content of 46% it has the lowest loss ratio of 0.0122.

4. Results

Application of methods to improve the efficiency packing cigarette. After applied the results to the production of packing process cigarette 2 between July and August 2017. Found that (1) organization and labor have study and improve the structure organizations and labor, by adopting a formal organizational chart or writing. To have a clear flow of work order, and set the scope of duties or job characteristics to do of each position clearly in order to control work monitoring and evaluation as well as educate staff by training. The cigarette packer did not receive the load because it was waiting to be transported to ten packets it decreases from 765 minutes were 167.5 minutes or decreased

78.11%. (2) Machine used TPM technique to increase productivity can be summarized after the improve as the following, The average monthly loss of the packing machine was reduced from 1,117.5 minutes to 587.5 minutes, or a decrease of 47.43%, the average time of loss per month of packing machine was reduced from 2,163 minutes to 2,008 minutes or 7.17%. (3) Raw material can be summarized after the improve as the following: time lost with the packing machine does not get the work load due to raw materials defects and package seams was broken, the monthly average decreased from 672.5 minutes to 91.5 minutes decreased by 86.39%. And the proportion of loss package seams was broken average monthly decreased from 0.8347 to 0.2177, a decrease of 0.617 or decreased 73.92%, the overall loss content of cigarette packaging and roll production process 2 decreased from 0.0597 was 0.0111, decreasing 0.0486.

5. Conclusion & Recommendations

Conclusion the study result by use the Overall Machine Effectiveness (OEE) measure to improve efficacy and increase the productivity of the process packing cigarette 2 of factory number 5. Index of indicators in this study include Loading of machine speed, efficacy of machine and the quality rate has increased. As well as the overall effectiveness of the machine has also increased when compared with the index value before improve. It was found that the average value of Loading efficiency increased by 17.35% with a value of 92.59%. Machine efficiency increased by an average of 12.03% to 84.17%. The quality rate increased on average by 4.83%, to 98.88%, and the overall effectiveness of the machine has increased by an average of 26.02% the value is 77.07%. by adding these index. As a result, production efficiency increased from 52.27% to 79.66% and productivity increased from 9,899 packs per hour to increase to 15,058 packs per hour or increased to 52.12% of the original yield. the method applied to optimize the production process of packaging cigarettes can be reduce the losses. And increase the efficiency of the production process of packaging cigarettes is higher than 75% according to the target. In order to applied this approach production process of packaging cigarettes of other units for further efficiency. Should be study to increase the high efficiency of the machine. Because the researcher also sees that there are ways to increase the efficiency of the packing machine by reducing the loss of capacity of the time machine. Which is as high as 2,008 minutes.

References

[1] Wanchai Rijiravanich 2000 Productivity in the industry (Bangkok: Chulalongkorn University)
[2] Chitra Rukijkanpanich 2001 Management of maintenance (Bangkok: Chulalongkorn University)
[3] Dal B, Tug well P and Greatbanks R 2000 Overall equipment effectiveness as a measure of operation Improvement *International Journal and Production Management*, 20(12), pp 1488–1502
[4] Douglas C Montgomery et al 2011 *Inc* Statistics for Engineering (N.Y.: John Wiley & Sons)
[5] Parames Chutima 2004 The experimental design engineering (Bangkok: Chulalongkorn University)
[6] Tejaskumar S Parsana and Mihir T Patel 2014 A Case Study: A Process FMEA Tool to Enhance Quality and Efficiency of Manufacturing Industry Bonfring *International Journal of Industrial Engineering and Management Science* vol 4, no 3 pp 145-152
[7] Bisgaard S and Fuller H T 1995a sample Size Estimates for 2k-p Design with Binary Response *Journal of Quality Technology* vol 27, no 4 pp 344-354
[8] Douglas C Montgomery 2005 *Inc* Design and Analysis of Experiments vol 6, ed (The United States of America: John Wiley & Sons)

Adaptive surrounding control for linear multi-agent systems with unknown disturbance

A Sharghi, M Baradarannia, F Hashemzadeh

Department of Control Engineering, Faculty of Electrical and Computer Engineering, University of Tabriz, Tabriz, Iran

E-mail: mbaradaran@tabrizu.ac.ir

Abstract. This paper deals with the surrounding control problem of multi-agent systems with linear non-identical dynamics for each agent. In these dynamics, a non-identical time-varying term is considered for each agent. It is assumed that, there exist unknown parameter vectors in time-varying terms of the followers. First, an estimator is introduced to estimate the geometric center of the leaders and then, an adaptive law is proposed to compensate the unknown parameter vector of the followers. Using Lyapunov stability theory, the stability of the system with the proposed control law is studied. Finally, a simulation example is given to show the effectiveness of theoretical results.

1. Introduction

In the last decades, cooperative control of network systems or multi-agent systems have been attracted lot of attentions [1]. Multiplicity of agents in these systems has created many applications in different domains such as formation of robotic systems [2], spacecraft formation flying [3], target tracking in sensor networks [4] and etc. Among the applications of cooperative control, the enclosing control has attracted many attentions. In the enclosing problem, a group of agents (called followers) surround other agents or a single agent (called leaders) in a certain geometric configuration. For achieving this purpose, many control methods have been used. For example, the authors in [5] propose cyclic pursuit strategies to achieve cooperative arresting for a single stationary target. In this work, the authors use consensus-based techniques to achieve cooperative enclosing for a moving leader. The authors in [6] utilize sliding mode method with incomplete leader information to build a surrounding control protocol, wherein a single moving target has been considered. Adaptive control method is proposed in [7] and [8] to accomplish enclosing formation for a leader when there exists uncertainty in the system. Also, [9] and [10] implement simultaneously the estimator and control strategy to accomplish enclosing control for targets when the system involves uncertainty or imperfect information.

In the surrounding control problem, all of the followers must encompass dynamic or static leaders. For this goal, the followers need geometric center of the leaders. When the geometric center is not available, an estimator is needed to be built. In the current paper, the surrounding control problem for a non-identical linear multi-agent system with a nonlinear term under undirected graph topology is investigated. Also, it is supposed that the nonlinear term of the followers consists of an unknown parameter vector. For this purpose, a distributed surrounding control protocol with adaptive law is proposed.

Content from this work may be used under the terms of the Creative Commons Attribution 3.0 licence. Any further distribution of this work must maintain attribution to the author(s) and the title of the work, journal citation and DOI.
Published under licence by IOP Publishing Ltd

IOP Conf. Series: Materials Science and Engineering **361** (2018) 012024 doi:10.1088/1757-899X/361/1/012024

The rest of this paper is organized as follows: In section 2 some preliminaries are given. Section 3 presents problem formulation. In Section 4, the main result and the center estimator of the leaders are established. Numerical simulation is provided in Section 5. Finally, the conclusions are drawn in section 6.

2. Preliminaries

In this section, some basic graph concepts are introduced. This paper only considers undirected graphs and does not assume any self-edges. For the graph $\mathcal{G} = \{\mathcal{V}, \mathcal{E}, \mathcal{A}\}$, $\mathcal{V} = \{v_i, i = 1, \dots, N\}$ means a set of nodes and $\mathcal{E} \subseteq \mathcal{V} \times \mathcal{V}$ means a set of edges. An edge denoted as $(i,j) \in \mathcal{E}$ means that agent j has access to the information of agent i and also, agent i is a neighbour of agent j. In this paper, the set of neighbours of agent i is represented by $\mathcal{N}_i$. Graph $\mathcal{G}$ is said to be undirected if $(i,j) \in \mathcal{E}$ implies $(j,i) \in \mathcal{E}$. An undirected graph is connected if there exists a path, i.e., a sequence of distinct edges such that consecutive edges are joint between any two vertices. The Adjacency matrix of the graph $\mathcal{G}$ is $\mathcal{A} = [a_{ij}] \in \mathbb{R}^{N\times N}$ with $a_{ij} = 1$ if the ith agent can receive information from jth agent, otherwise $a_{ij} = 0$. Throughout this paper, it is assumed that $a_{ii} = 0$ and the topology is fixed. The graph Laplacian matrix is $\mathcal{L} = [l_{ij}] = \mathcal{D} - \mathcal{A}$ where $\mathcal{D} = \text{diag}\{d_1, \dots, d_N\}$ is the degree matrix, where $d_i = \sum_{j=1}^{N} a_{ij}$ for $i = 1, \dots, N$. For an undirected graph, the Adjacency and Laplacian matrices are symmetric.

In this paper, a multi-agent system with N followers and M leaders is considered. It is supposed that each leader communicates with at least one follower. In this paper, the nodes set of the followers and the leaders are shown by V_F and V_L, respectively. The interconnection relationship between each follower and corresponding leader is indicated with the *explore* matrix and shown by $B = [b_{ik}] \in \mathbb{R}^{N\times M}$. The elements of this matrix is defined as $b_{ik} = 1$ if ith follower communicates with kth leader and $b_{ik} = 0$, otherwise. Kronecker product is indicated by $\otimes$.

3. Problem formulation

Consider a multi-agent system with N followers and M leaders. Dynamics of the followers and leaders are heterogeneous and described by

$$\dot{\boldsymbol{x}}_i(t) = -A_i\boldsymbol{x}_i(t) + B_i\boldsymbol{d}_i(t) + \boldsymbol{u}_i(t), \quad i \in V_F \tag{1}$$

where $\boldsymbol{x}_i(t)$ and $\boldsymbol{u}_i(t) \in \mathbb{R}^2$ are the state and control input of the ith follower, respectively. $\boldsymbol{d}_i(t)$ is an unknown time-varying disturbance of the ith follower and is formulated as $\boldsymbol{d}_i(t) = \varphi_i(t)\boldsymbol{\theta}_i, \quad i \in V_F$ in which $\varphi_i(t) \in \mathbb{R}^{2\times 2}$ is a known matrix function and $\boldsymbol{\theta}_i \in \mathbb{R}^{2\times 1}$ is an unknown constant parameter vector. A_i and B_i are constant matrices with compatible dimensions. It is assumed that the leaders move in $\mathbb{R}^2$ with dynamics

$$\dot{\boldsymbol{x}}_k(t) = -A_k\boldsymbol{x}_k(t) + B_k\boldsymbol{u}_k(t), \quad k \in V_L \tag{2}$$

where $\boldsymbol{x}_k(t) \in \mathbb{R}^2$ is the state of the kth leader and $\boldsymbol{u}_k(t) \in \mathbb{R}^2$ is the control input of the leader which is unknown for the followers. A_k and B_k are constant matrices with compatible dimensions. The dynamics equations (1)-(3) have been adopted from [11]. Throughout the paper, we make the following assumption:

Assumption 1: The movement of each leader around the center of the leaders is bounded. That is, there exists a real positive number $\mathcal{M} < \infty$ such that $\mathcal{M} = \sup\{|\boldsymbol{x}_i(t) - \overline{\boldsymbol{r}}(t)|, \ i = 1,2, \dots, N\}$. Here, $\overline{\boldsymbol{r}}(t)$ denotes the geometric center of the leaders and $\overline{\boldsymbol{r}}(t) = \frac{1}{M}\sum_{k=1}^{M}\boldsymbol{x}_k(t)$. The goal is to find a distributed adaptive control input for the followers which ensures creation of a certain geometrical configuration around the leaders, where the number of followers is allowed to be different from the number of leaders. To achieve this purpose, we should have

$$\lim_{t\to\infty}(\boldsymbol{x}_i(t) - \overline{\boldsymbol{r}}(t)) = \rho([\cos\left(\tfrac{i.2\pi}{N}\right) \quad \sin\left(\tfrac{i.2\pi}{N}\right)]^T) \tag{3}$$

where $\rho > \mathcal{M}$.

IOP Conf. Series: Materials Science and Engineering **361** (2018) 012024 doi:10.1088/1757-899X/361/1/012024

4. Main result

Suppose that the unknown nonlinear time-varying disturbance of the followers, i.e. $\boldsymbol{d}_i(t)$ is parameterized as $\boldsymbol{d}_i(t) = \varphi_i(t)\boldsymbol{\theta}_i, \quad i \in V_F$, where $\varphi_i(t) \in \mathbb{R}^{2\times 2}$ is a uniformly bounded function and $\boldsymbol{\theta}_i \in \mathbb{R}^{2\times 1}$ is an unknown constant parameter vector.

Because the unknown parameter vector $\boldsymbol{\theta}_i$ is not available to any follower, the ith follower estimates the unknown vector $\boldsymbol{\theta}_i$ by $\widehat{\boldsymbol{\theta}}_i$. Then, the estimation of $\boldsymbol{d}_i(t)$ by the ith follower is expressed as $\widehat{\boldsymbol{d}}_i(t) = \varphi_i(t)\widehat{\boldsymbol{\theta}}_i$.

The local surrounding error for ith follower is defined as

$$\boldsymbol{e}_i(t) = \boldsymbol{x}_i(t) - \bar{\boldsymbol{r}}(t) - \rho(\left[\cos\left(\tfrac{i.2\pi}{N}\right) \quad \sin\left(\tfrac{i.2\pi}{N}\right)\right]^T) \tag{4}$$

and global error vector is expressed as

$$\boldsymbol{e}(t) = \boldsymbol{x}(t) - \boldsymbol{1}_N \otimes \bar{\boldsymbol{r}}(t) - \rho(\boldsymbol{1}_N \otimes \left[\cos\left(\tfrac{i.2\pi}{N}\right) \quad \sin\left(\tfrac{i.2\pi}{N}\right)\right]^T) \tag{5}$$

Here, $\boldsymbol{1}_N$ represents a column vector and it's all elements are 1. It is obvious that the surrounding control is achieved when $\lim_{t\to\infty} \boldsymbol{e}(t) \to 0$.

As it can be seen, to achieve certain geometrical configuration, the geometric center of leaders should be available for each follower. In general, each follower has access to only a subset of the leaders. Therefore, a decentralized estimator is constructed for each follower to estimate the geometric center of the leaders. Then, the estimated center is used to design the controller.

4.1. *Estimator for the center of Leaders*

To perform a certain geometrical configuration, we introduce the following estimator for the center of leaders.

$$\widehat{\boldsymbol{x}}_i(t) = \phi_i(t) + \tilde{\boldsymbol{l}}_i(t), \quad i \in V_F \tag{6}$$

Here, $\widehat{\boldsymbol{x}}_i(t)$ is the estimation of the center of leaders for the ith follower and $\phi_i(t)$ is a dynamic variable with the following dynamics.

$$\dot{\phi}_i(t) = \alpha \sum_{k=1}^{N} a_{ik} \left[\frac{\widehat{\boldsymbol{x}}_k(t) - \widehat{\boldsymbol{x}}_i(t)}{\|\widehat{\boldsymbol{x}}_k(t) - \widehat{\boldsymbol{x}}_i(t)\|}\right] \tag{7}$$

where $\phi_i(0) = 0$, $\alpha > 0$ is a constant, and $\tilde{\boldsymbol{l}}_i(t)$ is defined as

$$\begin{bmatrix} \tilde{\boldsymbol{l}}_1(t) \\ \tilde{\boldsymbol{l}}_2(t) \\ \vdots \\ \tilde{\boldsymbol{l}}_N(t) \end{bmatrix} = \frac{N}{M}\left(I_n \otimes \begin{bmatrix} \frac{b_{11}}{\sum_{k=1}^{N} b_{k1}} & \frac{b_{12}}{\sum_{k=1}^{N} b_{k2}} & \cdots & \frac{b_{1M}}{\sum_{k=1}^{N} b_{kM}} \\ \frac{b_{21}}{\sum_{k=1}^{N} b_{k1}} & \frac{b_{22}}{\sum_{k=1}^{N} b_{k2}} & \cdots & \frac{b_{2M}}{\sum_{k=1}^{N} b_{kM}} \\ \vdots & \vdots & \ddots & \vdots \\ \frac{b_{N1}}{\sum_{k=1}^{N} b_{k1}} & \frac{b_{N2}}{\sum_{k=1}^{N} b_{k2}} & \cdots & \frac{b_{NM}}{\sum_{k=1}^{N} b_{kM}} \end{bmatrix}\right) \begin{bmatrix} \boldsymbol{x}_1(t) \\ \boldsymbol{x}_2(t) \\ \vdots \\ \boldsymbol{x}_M(t) \end{bmatrix} \tag{8}$$

in which $\boldsymbol{x}_j(t)$ denotes the position of the leaders.

In [12], it has been shown that the estimator defined by (6)-(8) converges to the leaders' center in a finite time T. Thus

$$\widehat{\boldsymbol{x}}_i(t) = \frac{1}{M}\sum_{k=1}^{M} \boldsymbol{x}_k(t), \quad t > T = \frac{1}{\delta}\sqrt{V(0)} \tag{9}$$

where $\delta \triangleq \frac{\sqrt{2N}[c_1-(N-1)c_2]}{2(N-1)}$ with $c_1 > c_2$, wherein c_1 and c_2 are real positive constants. For more details, one can refer to [13].

From (9), it can be seen that $\widehat{\boldsymbol{x}}_i(t) = \bar{\boldsymbol{r}}(t)$ when $t > T$. Then, the following modification is made for the local surrounding error of ith follower.

$$\boldsymbol{e}_i(t) = \boldsymbol{x}_i(t) - \widehat{\boldsymbol{x}}_i(t) - \rho(\left[\cos\left(\tfrac{i.2\pi}{N}\right) \quad \sin\left(\tfrac{i.2\pi}{N}\right)\right]^T) \tag{10}$$

Also, the global error vector is changed to:

IOP Conf. Series: Materials Science and Engineering **361** (2018) 012024 doi:10.1088/1757-899X/361/1/012024

$$\boldsymbol{e}(t) = \boldsymbol{x}(t) - \widehat{\boldsymbol{x}}(t) - \rho(\mathbf{1}_{\boldsymbol{N}} \otimes \left[\cos\left(\frac{i.2\pi}{N}\right) \quad \sin\left(\frac{i.2\pi}{N}\right)\right]^T) \tag{11}$$

where $\widehat{\boldsymbol{x}}(t) = [\widehat{\boldsymbol{x}}_1(t) \quad \widehat{\boldsymbol{x}}_2(t) \quad \cdots \quad \widehat{\boldsymbol{x}}_N(t)]^T$.

By defining $\boldsymbol{e}(t)$ as (11), if $\lim_{t\to\infty} \boldsymbol{e}(t) \to 0$ the surrounding control problem is solved asymptotically.

4.2. *Surrounding Control Protocol*

In this subsection, we propose a surrounding control strategy which achieves control objective (3) for any initial position of the followers and leaders. With considering the decentralized center estimator (6)-(8), the adaptive control protocol and decentralized adaptive laws for ith follower are proposed as follows.

$$\boldsymbol{u}_i(t) = \rho A_i(\mathbf{1}_{\boldsymbol{N}} \otimes \left[\cos\left(\frac{i.2\pi}{N}\right) \quad \sin\left(\frac{i.2\pi}{N}\right)\right]^T) - cK\boldsymbol{e}_i(t) + \dot{\widehat{\boldsymbol{x}}}_i(t) - B_i\varphi_i(t)\widehat{\boldsymbol{\theta}}_i + A_i\widehat{\boldsymbol{x}}_i(t) \tag{12}$$

$$\dot{\widehat{\boldsymbol{\theta}}}_i^{\,T}(t) = \boldsymbol{e}_i^{\,T}(t)P_i^{-1}B_i\varphi_i(t) \tag{13}$$

In (12), $c > 0$ is a constant number, $\dot{\widehat{\boldsymbol{x}}}_i(t)$ denotes the velocity of estimation of the center of leaders for the ith follower, $K_i = \frac{1}{2}P_i^{-1}$, and P_i is a positive definite solution to the following linear matrix inequality (LMI).

$$A_iP_i + P_iA_i^T - cI > 0 \tag{14}$$

Equations (12) and (13) can be written in vector form as

$$\begin{aligned}\boldsymbol{u}(t) = &-c\mathrm{diag}(K_1 \quad K_2 \quad \cdots \quad K_N)\boldsymbol{e}(t) + \dot{\widehat{\boldsymbol{x}}}(t) + \mathrm{diag}(A_1 \quad A_2 \quad \cdots \quad A_N)\widehat{\boldsymbol{x}}(t)\\ &-\mathrm{diag}(B_1 \quad B_2 \quad \cdots \quad B_N)\Phi(t)\widehat{\boldsymbol{\Theta}}(t) + \rho\mathrm{diag}(A_1 \quad A_2 \quad \cdots \quad A_N)(\mathbf{1}_{\boldsymbol{N}} \otimes \left[\cos\left(\frac{i.2\pi}{N}\right) \quad \sin\left(\frac{i.2\pi}{N}\right)\right]^T)\end{aligned} \tag{15}$$

and

$$\dot{\widehat{\boldsymbol{\Theta}}}^T(t) = \boldsymbol{e}^T(t)\mathrm{diag}(P_1^{-1} \quad P_2^{-1} \quad \cdots \quad P_N^{-1})\,\mathrm{diag}(B_1 \quad B_2 \quad \cdots \quad B_N)\Phi(t) \tag{16}$$

where $\boldsymbol{e}(t) = [\boldsymbol{e}_1(t) \quad \boldsymbol{e}_2(t) \quad \cdots \quad \boldsymbol{e}_N(t)]^T$, $\boldsymbol{u}(t) = [\boldsymbol{u}_1(t) \quad \boldsymbol{u}_2(t) \quad \cdots \quad \boldsymbol{u}_N(t)]^T$, $\widehat{\boldsymbol{\Theta}}(t) = [\widehat{\boldsymbol{\theta}}_1 \quad \widehat{\boldsymbol{\theta}}_2 \quad \cdots \quad \widehat{\boldsymbol{\theta}}_N]$ and $\Phi(t) = \mathrm{diag}(\varphi_1(t) \quad \varphi_2(t) \quad \cdots \quad \varphi_N(t))$.

Theorem 1. Suppose that assumption 1 holds and $\boldsymbol{d}_i(t)$ and $\boldsymbol{u}_k(t)$ are uniformly bounded for all agents. Under control law (12) and the parameter adaptive law (13) with $K_i = \frac{1}{2}P_i^{-1}$, where $P_i > 0$ is a solution of the LMI (14), the followers described by (1) create a certain geometrical configuration around the leaders.

Proof. Before presenting the proof, we define $\widetilde{\boldsymbol{\Theta}}(t) = \widehat{\boldsymbol{\Theta}}(t) - \boldsymbol{\Theta}$. The derivative of the global error vector (11) is $\dot{\boldsymbol{e}}(t) = \dot{\boldsymbol{x}}(t) - \dot{\widehat{\boldsymbol{x}}}(t)$.

By substituting (1) into $\dot{\boldsymbol{e}}(t)$, the derivative of the global error vector is obtained as

$$\dot{\boldsymbol{e}}(t) = -\mathrm{diag}(A_1 \quad A_2 \quad \cdots \quad A_N)\boldsymbol{x}(t) + \mathrm{diag}(B_1 \quad B_2 \quad \cdots \quad B_N)\boldsymbol{d}(t) + \boldsymbol{u}(t) - \dot{\widehat{\boldsymbol{x}}}(t) \tag{17}$$

where $\boldsymbol{d}(t) = [\boldsymbol{d}_1(t) \quad \boldsymbol{d}_2(t) \quad \cdots \quad \boldsymbol{d}_N(t)]^T$ and $\boldsymbol{x}(t) = [\boldsymbol{x}_1(t) \quad \boldsymbol{x}_2(t) \quad \cdots \quad \boldsymbol{x}_N(t)]^T$. Now, consider a Lyapunov function candidate as follows.

$$V(t) = \frac{1}{2}\boldsymbol{e}^{\boldsymbol{T}}(t)\mathrm{diag}(P_1^{-1} \quad P_2^{-1} \quad \cdots \quad P_N^{-1})\boldsymbol{e}(t) + \frac{1}{2}\widetilde{\boldsymbol{\Theta}}^{\boldsymbol{T}}(t)\widetilde{\boldsymbol{\Theta}}(t) \tag{18}$$

The derivative of $V(t)$ gives

$$\dot{V}(t) = \boldsymbol{e}^{\boldsymbol{T}}(t)\mathrm{diag}(P_1^{-1} \quad P_2^{-1} \quad \cdots \quad P_N^{-1})\dot{\boldsymbol{e}}(t) + \dot{\widehat{\boldsymbol{\Theta}}}^{\boldsymbol{T}}(t)\widetilde{\boldsymbol{\Theta}}(t) \tag{19}$$

By substituting equation (17), we have

$$\begin{aligned}\dot{V}(t) = &\boldsymbol{e}^{\boldsymbol{T}}(t)\mathrm{diag}(P_1^{-1} \quad P_2^{-1} \quad \cdots \quad P_N^{-1})\{-\mathrm{diag}(A_1 \quad A_2 \quad \cdots \quad A_N)\boldsymbol{x}(t) + \boldsymbol{u}(t) - \dot{\widehat{\boldsymbol{x}}}(t)\\ &+\mathrm{diag}(B_1 \quad B_2 \quad \cdots \quad B_N)\boldsymbol{d}(t)\} + \dot{\widehat{\boldsymbol{\Theta}}}^{\boldsymbol{T}}(t)\widetilde{\boldsymbol{\Theta}}(t)\end{aligned} \tag{20}$$

From (15), one has

$$\dot{V}(t) = \boldsymbol{e}^T(t)\mathrm{diag}(P_1^{-1} \quad P_2^{-1} \quad \cdots \quad P_N^{-1})\{-\mathrm{diag}(A_1 \quad A_2 \quad \cdots \quad A_N)\boldsymbol{x}(t) - \dot{\hat{\boldsymbol{x}}}(t) + \mathrm{diag}(A_1 \quad A_2 \quad \cdots \quad A_N)\hat{\boldsymbol{x}}(t) - c\,\mathrm{diag}(K_1 \quad K_2 \quad \cdots \quad K_N)\boldsymbol{e}(t) + \dot{\hat{\boldsymbol{x}}}(t) + \rho\,\mathrm{diag}(A_1 \quad A_2 \quad \cdots \quad A_N)(\mathbf{1}_N \otimes \left[\cos\left(\frac{i.2\pi}{N}\right) \quad \sin\left(\frac{i.2\pi}{N}\right)\right]^T) + \mathrm{diag}(B_1 \quad B_2 \quad \cdots \quad B_N)\boldsymbol{d}(t)\} + \dot{\hat{\boldsymbol{\Theta}}}^T(t)\widetilde{\boldsymbol{\Theta}}(t) - \mathrm{diag}(B_1 \quad B_2 \quad \cdots \quad B_N)\Phi(t)\widehat{\boldsymbol{\Theta}}(t) \tag{21}$$

Using (11), it follows that

$$\dot{V}(t) = \boldsymbol{e}^T(t)\mathrm{diag}(P_1^{-1} \quad P_2^{-1} \quad \cdots \quad P_N^{-1})\{-\mathrm{diag}(A_1 \quad A_2 \quad \cdots \quad A_N)\boldsymbol{e}(t) - c\,\mathrm{diag}(K_1 \quad K_2 \quad \cdots \quad K_N)\boldsymbol{e}(t) - \mathrm{diag}(B_1 \quad B_2 \quad \cdots \quad B_N)\Phi(t)\widetilde{\boldsymbol{\Theta}}(t)\} + \dot{\hat{\boldsymbol{\Theta}}}^T(t)\widetilde{\boldsymbol{\Theta}}(t) \tag{22}$$

By using adaptive laws (13), one can write

$$\dot{V}(t) = -\boldsymbol{e}^T(t)\{\mathrm{diag}(P_1^{-1} \quad P_2^{-1} \quad \cdots \quad P_N^{-1})\,\mathrm{diag}(A_1 \quad A_2 \quad \cdots \quad A_N) - c\,\mathrm{diag}(K_1 \quad K_2 \quad \cdots \quad K_N)\}\boldsymbol{e}(t) \tag{23}$$

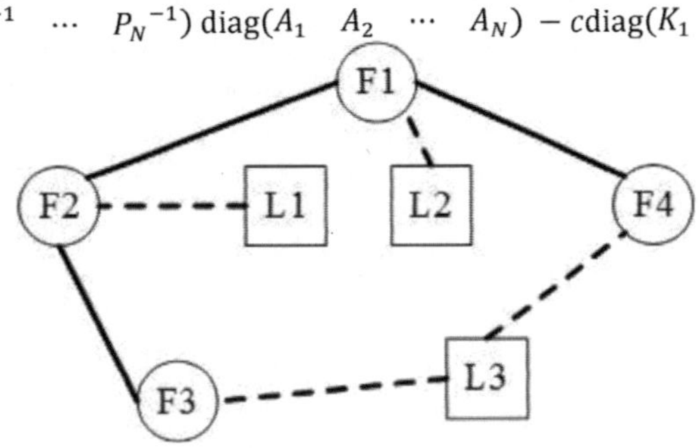

Figure 1. Communication topology

If we define $P_E = \mathrm{diag}(P_1^{-1} \quad P_2^{-1} \quad \cdots \quad P_N^{-1})$, $A_E = \mathrm{diag}(A_1 \quad A_2 \quad \cdots \quad A_N)$ and $K_E = \mathrm{diag}(K_1 \quad K_2 \quad \cdots \quad K_N)$, we can rewrite (23) as

$$\dot{V}(t) = -\frac{1}{2}\boldsymbol{e}^T(t)\{P_E A_E + A_E^T P_E - 2cP_E K_E\}\boldsymbol{e}(t) \tag{24}$$

Defining $K_E = \frac{1}{2}P_E$, multiplying both side with P_E^{-1} and using LMI (14), $\dot{V}(t)$ gets negative. $\boldsymbol{e}(t)$ and $\widetilde{\boldsymbol{\Theta}}(t)$ are uniformly bounded. The assumption on $\varphi_i(t)$ results that $\dot{\boldsymbol{e}}(t)$ in equation (22) is uniformly bounded. Because $\boldsymbol{e}(t) \in \mathcal{L}_2$ and $\dot{\boldsymbol{e}}(t) \in \mathcal{L}_\infty$, using Barbalat's lemma, we have $\lim_{t\to\infty} \boldsymbol{e}(t) \to 0$.

5. Simulation result

In this section, we present an example to evaluate our theoretical results. Consider a group of agents with four followers, three leaders and $n = 2$. The network connections are given in Fig. 1. In this figure, to show the communication between the followers and leaders, we use solid line and dashed line, respectively. The constant matrices of the followers are selected as

$$A_1 = \begin{bmatrix} 1.5 & -1 \\ -1 & 2 \end{bmatrix}, A_2 = \begin{bmatrix} 1 & 0 \\ 0 & 2 \end{bmatrix}, A_3 = \begin{bmatrix} 1 & -2 \\ 0 & 3 \end{bmatrix}, A_4 = 2I_2$$
$$B_1 = \begin{bmatrix} 1.5 & -0.1 \\ -0.2 & 2 \end{bmatrix}, B_2 = 2I_2, B_3 = \begin{bmatrix} -2 & -0.1 \\ -0.1 & -3 \end{bmatrix}, B_4 = \begin{bmatrix} -2 & 0 \\ -2 & -3 \end{bmatrix}. \tag{25}$$

The unknown nonlinear disturbance dynamics of the followers are chosen as: $\varphi_1(t) = \mathrm{diag}(0.5\cos(4t), e^{-2t} - \sin(2t))\boldsymbol{\theta}_1$, $\varphi_2(t) = \mathrm{diag}(0.9e^{-t}, -2)\boldsymbol{\theta}_2$, $\varphi_3(t) = \mathrm{diag}(1, -t)\,\boldsymbol{\theta}_3$, $\varphi_4(t) = \mathrm{diag}(0.5t, 0.9\cos(3t))\boldsymbol{\theta}_4$. The unknown parameters $\boldsymbol{\theta}_i$ are chosen as: $\boldsymbol{\theta}_1 = [1 \quad 0.5]^T$, $\boldsymbol{\theta}_2 = [-0.9 \quad 0.6]^T$, $\boldsymbol{\theta}_3 = [0.07 \quad 0.9]^T$, $\boldsymbol{\theta}_4 = [0.98 \quad -1.5]^T$.
The time-varying control input of the leaders are selected as follows.

$$A_1 = 2.5I_2, A_2 = \begin{bmatrix} 2 & 1 \\ 2 & 3 \end{bmatrix}, A_3 = \begin{bmatrix} 4 & 1 \\ 1 & 3 \end{bmatrix}, B_1 = \begin{bmatrix} -2 & -1 \\ -2 & 3 \end{bmatrix},$$

$$B_2 = \begin{bmatrix} -1 & 0 \\ 0 & -2 \end{bmatrix}, B_3 = \begin{bmatrix} -1 & 0 \\ 2 & 2 \end{bmatrix}. \quad (26)$$

The leader's constant matrices are selected as $\boldsymbol{u}_1(t) = [5 - 0.2t \quad 17 + 0.5t]^T$, $\boldsymbol{u}_2(t) = [4 \quad 7 + 0.1t]^T$ and $\boldsymbol{u}_3(t) = [2.25 \quad -0.3t]^T$. From (19) and (20) with $c = 2, \rho = 5\mathcal{M}$. Also, $\alpha = 9$ in the leaders center estimator (10)-(12).

The simulation results are presented in Fig. 2 and Fig. 3. In Fig. 2 the trajectories of the followers and leaders are depicted. In this figure, the surrounding of the followers around the leaders are shown with dashed line for better illustrations. It is clear that four followers enclose three leaders during the control process. Also, by enlarging the moving radius of the leaders, the moving radius of the followers increase to surround the leaders. Fig. 3 shows the surrounding error of the followers. It is clear that all of the trajectory errors tend to zeros as time increases.

Remark. If the goal is faster convergence of the global surrounding error vector reaching zero at shorter time, then it is necessary to change the control input to achieve finite-time stability or fixed-time stability. The authors suggest this as a future work.

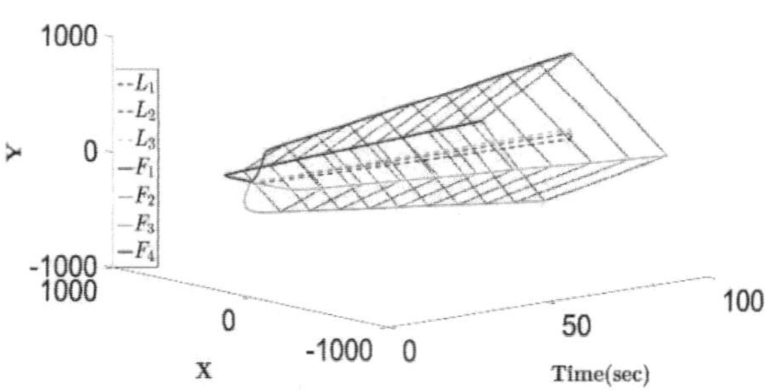

Figure 2. Trajectory of all agents

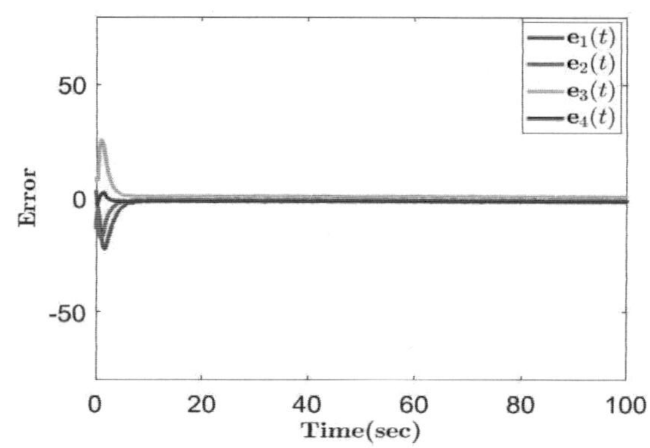

Figure 3. The global surrounding error vector

6. Conclusion

This paper addressed the adaptive surrounding control problem for non-identical linear multi-agent systems with nonlinear non-identical time-varying terms. It was demonstrated that the proposed method made the followers efficiently enclose the moving leaders in the presence of nonlinear unknown terms in the followers' dynamics. Numerical simulation revealed the efficacy of the results.

References

[1] Y Cao, W Yu, W Ren and G Chen, "An overview of recent progress in thestudy of distributed multi-agent coordination," *IEEE Transactions on Industrial Informatics*, 9, 427–438, 2013.

[2] J Desai, J Ostrowski and V Kumar, "Modelling and control of formations of nonholonomic mobile robots," *IEEE Transactions on Robotics and Automation*, 17, 905 –908, 2001.

[3] Y Xu, "Dynamics and control for formation flying system," PhD thesis, 2003, University of Florida.

[4] J Hu and X Hu, "Optimal target trajectory estimation and filtering using networked sensors," *Journal of Systems Science and Complexity*, 21, 3, 325–336, 2008.

[5] T H Kim, S Hara, and Y Hori, "Cooperative control of multi-agent dynamical systems in target-enclosing operations using cyclic pursuit strategy," *International Journal of Control*, 83, , 2040–2052, 2010.

[6] H Kawakami and T Namerikawa, "Cooperative target capturing strategy for multi-vehicle systems with dynamic network topology," *Proceeding of American Control Conference*, 2009, 635–640, 2009.

[7] M Kothari, R Sharma, I Postlethwaite, R W Beard and D Pack, "Cooperative target-capturing with incomplete target information," *Journal of Intelligent & Robotic Systems*, 72, no. 3-4, pp. 373–384, 2013.

[8] K Sato and N Maeda, "Target-enclosing strategies for multi-agent using adaptive control strategy," *Proceedings of IEEE Multi-Conference on Systems and Control*, 1761–1766, 2010.

[9] J Guo, G Yan, and Z Lin, "Local control strategy for moving target enclosing under dynamically changing network topology," *Systems and Control Letters*, 59, 10, 654–661, 2010.

[10] A Franchi, P Stegagno, and G Oriolo, "Decentralized multirobot encirclement of a 3D target with guaranteed collision avoidance," *Autonomous Robots*, 2015, doi: 10.1007/s10514-015-9450-3.

[11] J Sun and Z Geng, "Adaptive consensus tracking for linear multi-agent systems with heterogeneous unknown nonlinear dynamics," *International Journal of Robust and Nonlinear Control*, 26, 154-173, 2016.

[12] Y J Shi, R Li and T T Wei, "Target-enclosing control for second-order multi-agent systems," International Journal of Control, 46, 2279-2286, 2013.

[13] Y J Shi, R Li and K L Teo, "Cooperative enclosing control for multiple moving targets by a group of agents," *International Journal of Control*, 88, 80-89, 2015.

IOP Conf. Series: Materials Science and Engineering **361** (2018) 012025 doi:10.1088/1757-899X/361/1/012025

Airflow and Temperature Simulation in a Big Cleanroom to Reduce Contamination in an HDD Manufacturing Factory

Jiranan Naosungnoen[1] and Jatuporn Thongsri[1]

[1]Computer Simulation in Engineering Research Group, College of Advanced Manufacturing Innovation, King Mongkut's Institute of Technology Ladkrabang, Bangkok, Thailand

Email: Jatuporn.th@kmitl.ac.th

Abstract. Contamination is an important factor that must be well controlled at a stringent level of cleanliness. If it exceeds factory's specifications it shall contribute to poor quality products and product loss in the hard disk drive (HDD) manufacturing factory. This research used a CFD software, FLUENT to simulate the airflow pattern and temperature distribution in a big clean room to find the cause(s) of high contamination. The results of simulation revealed that there are areas in the production with vortex flow which caused high contamination similar to the actual measurement in that clean room. High air velocity that came out of the ULPA Filter at the ceiling above collided at the front of one Return grill nearby. It also caused high velocity vortex and high contamination in that area. To reduce the contamination, we closed one of the Return Grilled near that problematic area which could help reduce the vortex flow and the airflow pattern to be smoother than the previous clean room. Also, the contamination result showed a significant reduction in particle count. The temperature distribution after improvement action taken was unaffected and remained under the control. This research can help enhance a product quality and improve an efficacy of HDD manufacturing process

1. Introduction

Currently, all electronic devices or components are very small sized built in manufacturing in many industries. Small components need to be environmentally controlled in the production line, especially in the clean room area where the levels of cleanliness, temperature, and humidity must be controlled to ensure good quality for products in factory. Contamination is the one of factors controlled in the clean room in order to ensure the quality of the products to be up to the specifications and good per spec requirement. The experiment to study some complex of building and structure for understanding the whole production line is difficult. Therefore the one method that has been currently and successfully used is the Computational fluid dynamics (CFD) which is the method to simulate and validate our model easily. Many researches such as K. C.Noh et al. [1] successfully used CFD analysis based on particle concentration measurement and airflow by the study was carried out to look for the source of contamination and examine the route of contaminant transfer in the mini-environment applied in the LCD process clean room to improve results. J. Thongsri et al. [2-3] used CFD to simulate airflow and particle trace over and around a piece of hard disk drive (HDD) machine in factory to find the best airflow setup which was later deployed to the manufacturing process. The cause of the high particle counts was improper air-speed from the FFUs. Air speed was optimized to be in the suitable range which the airflow would block out nearby airborne particles and purge away particles generated by the

Content from this work may be used under the terms of the Creative Commons Attribution 3.0 licence. Any further distribution of this work must maintain attribution to the author(s) and the title of the work, journal citation and DOI.
Published under licence by IOP Publishing Ltd

machine effectively. The temperature is the one factor studied to control the clean room such as **J.** Thongsri [4] used CFD to study the condensation in the factory which the recommendation from the study can be used in the factory to help save costs in the factory as well. CFD was also applied to the medical field such as in hospital rooms to reduce contamination in an operating room and to optimize rooms for efficient ventilation, as opening and closing doors can transfer contamination into other rooms in a hospital [5-7]. CFD is a fluid mechanics software that uses numerical analysis and algorithms to solve problems that involve fluid flows and display the solution data in colour graphics which makes it easy to analyse the model. The advantages of using CFD mainly are less time consumption and cost saving compared to conventional experimental methods that may be difficult to perform. Still, the accuracy of the results may be questionable. This research study HDD Manufacturing AAB clean room at Seagate Technology, Thailand (Korat plant), which manufactured HDD reader head. This paper studied the airflow pattern and temperature distribution in the big clean room which is scheduled for an environment big cleaning day every year. The environment will change after the big clean due to changes in the inlet and outlet. The scope of this study covers airflow pattern, temperature distribution and particle count by using experiment and ANSYS FLUENT. The purpose is to find the way to reduce contamination based on the experimental results. This research is different from the other research mentioned above as we consider the temperature of this clean room by using standard k-ε.

2. Theoretical

ANSYS Fluent [8] was used in this study by solving conservation equations that consist of mass, momentum and energy conservations, the equation for conservation of mass, or continuity equation, which can be written as follows,

$$\frac{\partial \rho}{\partial t}+\frac{\partial}{\partial x_i}(\rho u_i)=0. \tag{1}$$

The momentum and energy conservations can be written in (2) and (3)

$$\frac{\partial u_i}{\partial t}+\frac{\partial}{\partial x_j}\left(\rho u_i u_j\right)=-\frac{\partial P_i}{\partial x_i}+\mu\frac{\partial}{\partial x_j}\left(\frac{\partial u_i}{\partial x_j}+\frac{\partial u_j}{\partial x_i}\right)+\rho g_i, \tag{2}$$

$$\frac{\partial(peT)}{\partial t}+\frac{\partial}{\partial x_j}\left(peTu_j\right)= pq+\frac{\partial}{\partial x_j}\left(k\frac{\partial T}{\partial x_j}\right)-\frac{\partial(pu_j)}{\partial x_j}+\frac{\partial(\tau_{ij}u_i)}{\partial x_j}+pg_i \tag{3}$$

This study selected a standard k-ε model [9] which is a model based on model transport equations of turbulence kinetic energy (k) and dissipation rate (ε). The model transport equation for k derived from the exact equation, while the model transport equation for ε derived from physical reasoning and bares little resemblance to its exact mathematical counterpart. The standard k- ε model in ANSYS FLUENT falls within this class of models and has become the workhorse of practical engineering flow calculations in the time proposed by Launder and Spalding [9]. The equations of turbulence kinetic energy (k) and dissipation rate (ε) are written as follows,

$$\frac{\partial(\rho k)}{\partial x_j}+\frac{\partial(\rho k u_i)}{\partial x_i}=\frac{\partial}{\partial x_j}\left[\left(\mu+\frac{\mu_t}{\sigma_k}\right)\frac{\partial k}{\partial x_j}\right]G_k-\rho\varepsilon, \tag{4}$$

$$\frac{\partial(\rho\varepsilon)}{\partial t}+\frac{\partial(\rho u_i\varepsilon)}{\partial x_i}=\frac{\partial}{\partial x_j}\left(\left(\mu+\frac{u_t}{\sigma\varepsilon}\right)\frac{\partial \varepsilon}{\partial x_i}\right)+C_{\varepsilon 1}\frac{\varepsilon}{k}G_k-C_{\varepsilon 2}\rho\frac{\varepsilon^2}{k}, \tag{5}$$

where $G_k=-\overline{\rho u_i' u'_j}\frac{\partial u_j}{\partial x_i}$, represents the generation of turbulence kinetic energy due to the mean velocity gradients. The model constants have been established to ensure that the model performs well for certain canonical flows. The model constants are;

$$\mu_t=\rho C_\mu\frac{k^2}{\varepsilon},\quad C_\mu=0.09, C_{\varepsilon 1}=1.44, C_{\varepsilon 2}=1.92, \sigma_k=1.0,\ \sigma_\varepsilon=1.33$$

ANSYS FLUENT solved a set of partial differential equations of (1) - (5) to determine the airflow pattern, temperature distribution and shown in graphic colour.

3. Research Method

3.1. Model Description

Our research studied the clean room where the reader heads of hard disk drives are manufactured. The clean room is classified to be class 6 with particles of size 0.3 μm < 102,000 part/m^3 and particles of size 0.5 μm < 35,200 part/m^3 [10]. When particle counts exceeded the specification, or we called "Contamination out of control limit". The product quality would be poor or rejected. The level of cleanliness or contamination must be controlled in this clean room. Every year the factory will have a technical wipe down activity. The environment will be changed after the startup of the production line. The reason why we selected this clean room to study is to reduce or control contamination to be in the control level. Another purpose is to study the clean room whether it can be reused after an environmental change. The clean room layout shows in Fig.1. The environment was in operation during measuring all parameters such as air flow, pressure, temperature and particle count. The dimensions of this clean room were 26.81 m (L) × 27.1 m (W) × 3 m (H). There are 3 smaller "service rooms" inside the clean room which machine engines are stored. The air inside the clean room will flow out to these service rooms. On the top of clean room wall, there are 25 return grilles and on the ceiling are 55 ULPA Filters with the size of 0.6 m x 1.21 m. Outlets are Return grilled and the panel on the wall. Inlets are ULPA filters. These inlet and outlet are filled in the fluid model with different colour .We constructed models with different numbers of tetrahedron elements and nodes. All of these mesh models were put through a mesh analysis, and it was found that the mesh model was constructed with 10.36 million elements and 1.95 million nodes were the best model in terms of the quality of the results and computation time.

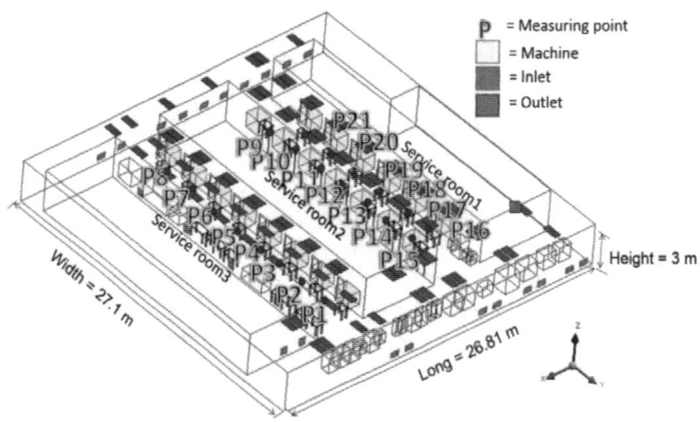

Figure 1. Clean room layout

3.2. ANSYS FLUENT Settings

There are 55 inlets (ULPA Filters) that are mass flow inlets of which boundary conditions around 0.1-1.4 kg/s and 47 outlets of which boundary conditions are 0.24% - 5.51%. All air velocity measurements are done with an airflow meter under actual operating conditions. The inlet velocities are measured under the ULPA filters at the center of each filter, and the outlet velocities are measured at the center of the return grille and the measuring time at each position is at least 10 s. Approximately 10 measurements are averaged at each point. The flow in the cleanroom is assumed to be steady. The standard k- ε transport equations of turbulence model are used for the simulation. Fluent settings are as shown in Table 1.

Table 1. Fluent settings.

Parameters	Settings	Parameters	Settings
Solver	Pressure based	**Algorithm**	Pseudo Transient
Velocity Formulation	Absolute	**Solution Initialization**	Hybrid Initialization
Time	Steady	**Convergence Criterion**	1.00E-06
Model	k-epsilon	**Thermal**	On
Pressure-Velocity Coupling	Coupled	**Energy**	On

4. Result and Discussion

To verify the model and results of simulation, we measured the air velocity in the points P1-P21 shown in Fig. 1 which are compared with the results of simulation. We selected air velocities points in the area where the production parts are placed to verify the model whether similar with the real environment compared with CFD simulation. During the measurements, the operators and the machine were operating normally in the clean room. The air flow meter was activated for 1 minute for each measuring point. The positions along the x-axis of P1~P8 was X=17.45 m, of P9~ P15 was X=8.49 m, and of P16~ P21 was X=7.50 m. The height Z from the ground to the point where the airflow velocity was measured was 0.82 m for each point. The comparison result shows that the percentage error (%) of every point was less than 15% as Table 2. We concluded that our model from CFD was able to accurately reproduce the air velocities in the clean room

Table 2. shows the velocities of airflow at P1~P21.[11]

Machine Zone 1 X= 17.45 m				
Point	Y (m)	velocity (m/s) Experiment	velocity (m/s) CFD	% Error
P1	21.3	0.219	0.206	5.89
P2	18.7	0.068	0.061	10.87
P3	16.2	0.134	0.143	7.03
P4	13.6	0.131	0.144	9.88
P5	11.1	0.030	0.029	5.19
P6	8.4	0.248	0.278	12.09
P7	5.8	0.338	0.344	1.78
P8	3.3	0.265	0.231	12.97

Machine Zone 2 X= 8.49 m				
Point	Y (m)	velocity (m/s) Experiment	velocity (m/s) CFD	% Error
P9	4.8	0.078	0.071	8.54
P10	7.3	0.089	0.084	6.31
P11	9.9	0.075	0.067	10.99
P12	12.5	0.067	0.060	9.66
P13	15.0	0.060	0.056	6.32
P14	16.9	0.064	0.060	5.55
P15	19.5	0.076	0.066	13.14

Machine Zone 3 X= 7.5 m				
Point	Y (m)	velocity (m/s) Experiment	velocity (m/s) CFD	% Error
P16	19.0	0.085	0.079	6.18
P17	16.3	0.071	0.070	1.39
P18	13.8	0.091	0.099	8.45
P19	11.2	0.258	0.237	8.02
P20	8.7	0.100	0.089	11.30
P21	6.2	0.091	0.100	9.37

We considered the graphic contour of velocities along the x-axis of P1~P8 was Plan X=17.45 m, of P9~ P15 was Plane X=8.49 m, and of P16~ P21 was Plan X=7.50 m where the production parts were placed and observed Plan X=17.45 m had higher velocity than other plan shown as Fig.2. This was probably shown as graphic colour result at P6-P8 higher air velocities than other measuring points. The observation is the air velocity flows out of the ULPA Filter at the ceiling above before colliding and returning to the grilled causing high velocity vortex near P6-P8.

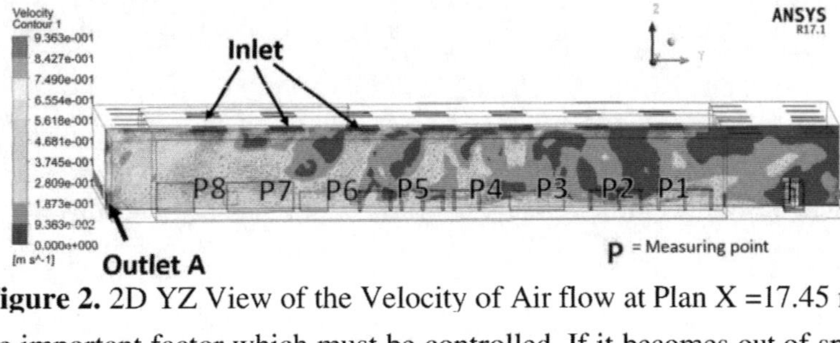

Figure 2. 2D YZ View of the Velocity of Air flow at Plan X =17.45 m

Temperature is one important factor which must be controlled. If it becomes out of specification or lower than the dew point, it may result in other problems such as condensation resulting in defects [4].

In this study, we simulated and measured the temperature distribution in the clean room. From the measurement and simulation, the temperature is around 19.7 °C – 22.3 °C which is within specification of this clean room by 18.3 °C – 23.9 °C. The temperature distribution in the Plane X=17.45 shows in Fig. 4 of next section (before action). At a center of clean room, the temperature is around 21.27 °C -23.9 °C but in the area near wall of clean room, the temperature is around 19.7 °C – 21.27 °C which is lower than that of the center.

4.1. Contamination control

We considered the contamination point to be the same as the velocity measurement in the previous section at points P1~P21 as shown in Fig.1. We used optical particle counter Lighthouse, Soliar 3100+ to measure particle for investigation. All measurements were measured under normal operating conditions. The particle counter was activated for 1 minute and 5 measurements were averaged at each point. We observed that the particle count at every point was within the specification except the point P6, P7 and P8 which were higher than those at other points related with the velocity data as shown in the previous section. We speculate that those point have higher particles from the vortex flow around that area from the simulation result. We considered to reduce the contamination at P8 which is highest.

We observed the P6, P7 and P8 from CFD simulation. The air velocity coming out of the ULPA Filter at the ceiling above P6, P7 and P8 were high and the airflow was not smooth due to air vortex at the in front of Outlet A. We were able to understand that the main contamination source near P8 was a vortex flow.

At first, we tried to reduce the vortex flow by adjusting the airflow rate coming out of the UPLA filter but it could not be done because the flow rate of all ULPA filters were centrally controlled and could not be adjusted individually. Moreover, the cabinet that obstructed around that area could not be moved away to another location. Therefore, we decided to close the Outlet A (flow rate = 0 kg/s) at the wall near these positions. The vortex flow and particle concentration were reduced from the measurement after the action was taken. We compared the before and after-action from CFD simulation in Fig.3. They showed that the airflow velocity at point P6, P7 and P8 improved significantly after the improvement. Also, the temperature after the action do not significantly impact nor change, therefore still landing within the specification as Fig.4. The study showed that after closing the one return grilled, this helped reduce vortex flow and did not impact the clean room temperature. This temperature simulation can be future improved for further study about this clean room since some reports showed this clean room was contaminated out of spec in the high temperature area in the simulation while some reports mentioned condensation problems on the machine around the lower temperature zone in the simulation area, related with the work of **J. Thongsri** [4]. If the temperature is lower than the dew point, water will condense. The research can help factory save costs from contaminated product, avoid water condensation problem and increase yield rate of the product.

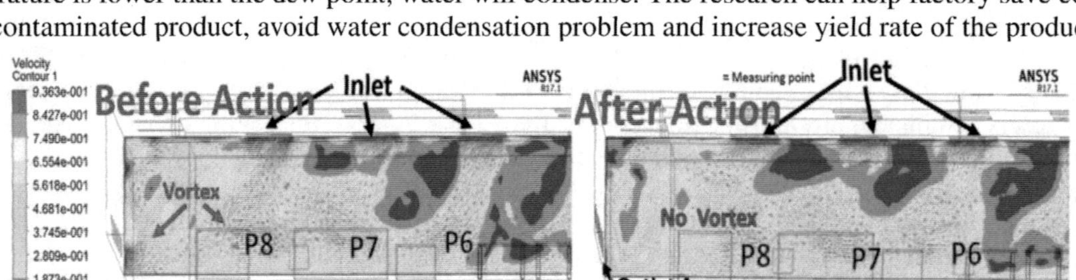

Figure 3. YZ View of the Velocity of Air flow at Plan X =17.45 m (Before- After Action).

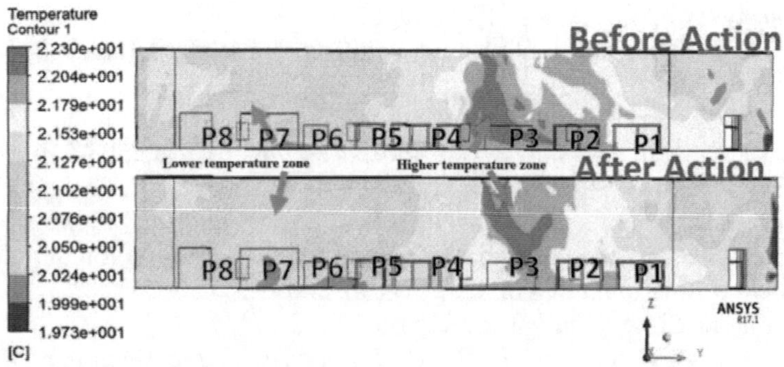

Figure 4. YZ View of the temperature After Action at Plan X =17.45 m (Before- After Action).

5. Conclusion

We studied the airflow inside a big clean room in a hard disk drive manufacturing factory in order to investigate the contamination, the airflow pattern, temperature distribution and particle count compared to an actual measurement. The CFD results show accurately simulated airflow pattern and high velocity with the vortex flow from return grilled which is related to the measurement data from the particle count. Areas of high particle which has vortex flow collecting the particles were found near the observed positions P8, P7 and P6 respectively. After the improvement action was taken by closing the Outlet A (one of return grilled) with flow rate = 0 kg/s at the wall near these positions, the vortex flow and contamination were reduced as Fig.5, and the temperature distribution of this clean room were left unaffected. This study confirms that CFD methodology can be utilized to predict the distribution of contaminated particles whenever an environmental change may take place. This model was submitted to the factory and was accepted to be used as information for clean room improvement

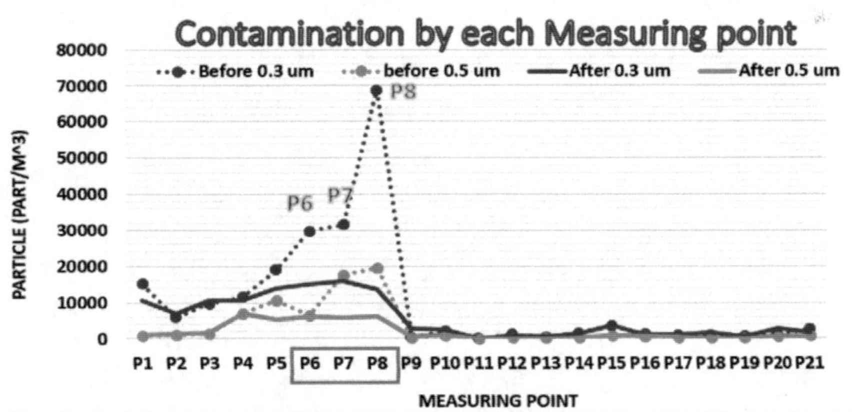

Figure 5. Contamination at P1-P21 Before-After the improvement

References

[1] Noh K C, Kim H S and Oh M D 2010 Study on contamination control in a minienvironment inside clean room for yield enhancement based on particle concentration measurement and airflow CFD simulation *Building and Environment* vol 45 pp 825-831

[2] Thongsri J and Pimsarn M 2015 Optimum airflow to reduce particle contamination inside welding automation machine of Hard Disk Drive production line *International journal of precision engineering and manufacturing* vol 16 no 3 pp 509-515

[3] Thongsri J 2017 A Problem of Particulate Contamination in an Automated Assembly Machine Successfully Solved by CFD and Simple Experiments *Mathematical Problems in Engineering* vol 2017 Article ID 6859852

[4] Thongsri J 2017 A Successful CFD-Based Solution to a Water Condensation Problem in a Hard Disk Drive Factory *IEEE Access* vol 5 pp 10795- 10804

[5] Sadrizadeh S and Holmberg S 2014 Surgical clothing systems in laminar airflow operating room: a numerical assessment *Infection and Public Health* vol 7 pp 508-516

[6] Me´ndez C San Jose´ J F Villafruela J M and Castro F 2008 Optimization of a hospital room by means of CFD for more efficient ventilation *Energy and Buildings* vol 40 pp 849–854

[7] Shiha Y C Chiua C C and Wangb O Dynamic airflow simulation within an isolation room *Building and Environment* vol 42 pp 3194–3209

[8] 2013 *ANSYS Fluent Theory Guide* U.S.A. p 46

[9] Launder B E and Spalding D B 1972 *Lectures in Mathematical Models of Turbulence* Academic Press England London

[10] 1999 Cleanrooms and associated controlled environments Part 1: Classification of air cleanliness, ISO 14644-1

[11] Naosungnoen J and Thongsri J 2018 Simulation of airflow in a cleanroom to solve contamination problem in an HDD production line *International Journal of Mechanical Engineering and Robotics* vol 7 pp 41

Acknowledgment

This work was financially sponsored by a contract no. MSD5810072 of Research and Researcher for Industries (RRI) under the Thailand Research Fund (TRF), and Seagate Technology (Thailand) Ltd.. Facilities were supported by College of Advanced Manufacturing Innovation (AMI) and King Mongkut's Institute of Technology Ladkrabang (KMITL).

IOP Conf. Series: Materials Science and Engineering **361** (2018) 012026 doi:10.1088/1757-899X/361/1/012026

A Method of Maneuver Detection Based on Multi-step Innovation

WANG Yong

Jiangsu Automation Research Institute, Lianyungang, 222061, Jiangsu Province, China.

E-mail: wyong_jari@aliyun.com

Abstract. One way to improve the processing capability for high-speed mobile anti-ship missiles is to improve the target data tracking rate, so that the Fire Control System can obtain more target information, but the improvement of the data sampling rate will reduce the signal - to - noise ratio of single - step window measurement by the traditional maneuver detection algorithm. In this paper, a maneuver detection algorithm based on multi - step prediction is proposed. The algorithm uses the current measurement value and the multi - step prediction value information, and uses the maneuver detection method based on attenuation memory. Through the simulation calculation method, the relationship between the minimum maneuver detection delay time and the attenuation coefficient, and the prediction step are given. At the same time, the method can also effectively reduce the detection of false alarm rate.

1. Introduction

Maneuvering target tracking mainly included the preprocessing of observation data, the maneuvering target model, the maneuvering detection and motion identification, the filtering and prediction, the tracking coordinate system and the selection of filter state variables which were location and description of target space. The difficulties mainly came from two aspects: the first difficulty was the uncertainty of the target movement, the second difficulty was the noise of observation data. In order to solve this problem, an important step was the maneuver detection[1,2,3].

The main idea of maneuver detection was to conduct a maneuver detection in accordance with a particular criterion or logic with the change of innovation vector $d(k) = \mathrm{Y(k)} - \mathrm{H}\hat{\mathrm{X}}(k/k-1)$ which came from observation data Y(k) and state prediction data $\mathrm{H}\hat{\mathrm{X}}(k/k-1)$. Based on maneuver detection results, we could adaptive adjust system state noise matrix, estimate and predict the motion state of maneuvering targets at last, achieve to tracking target. Therefore, the performance of maneuvering target tracking could be detected in a reliable and timely detection of target maneuver model.

In 21 century, due to the rapid development of electronic technology, the performance of high-speed target had been greatly improved. In order to improve the Terminal defense system processing capacity of supersonic anti-ship missiles, one way is to improve the fire control system data rate to fully use the tracker's target measurement data. In view of the improvement of system data rate, the traditional maneuver detection method[4,5,6,7] based on single-step innovation is no longer applicable. In this paper, a new method of maneuvering detection based on multi-step innovation was proposed, which can be used to extract the target maneuver information, which is convenient for maneuver detection.

Content from this work may be used under the terms of the Creative Commons Attribution 3.0 licence. Any further distribution of this work must maintain attribution to the author(s) and the title of the work, journal citation and DOI.
Published under licence by IOP Publishing Ltd

Through simulation calculation, for a given false alarm rate, the detection method gives the relationship between the minimum maneuver detection delay and the attenuation coefficient.

2. Multi-step innovation calculation method

The target dynamic model was used to describe the change of the target state X with the change of time. Linear state space model (discrete time model) was:

$$\begin{cases} X(k+1)=\Phi(k+1|k)X(k)+U(k)\overline{\alpha}+W(k); \\ Y(k)=H(k|k)X(k)+V(k); \end{cases} \tag{1}$$

Taking the discrete state equation as an example, X (k), Y (k) and U (k) were the vector of the target state, observation and control inputs at k time; Φ(k+1,k) and H (k) were the appropriate state transition matrix and observation matrix W (k) and V (k) were system model noise and observation noise, respectively , which were Gauss white noise sequence, Statistical characteristics were:

$$\begin{cases} E[W(k)]=0; E[W(k)W^{\tau}(j)]=Q\cdot\delta_{kj}; \\ E[V(k)]=0;\ E[V(k)V^{\tau}(j)]=R\cdot\delta_{kj}; \end{cases} \tag{2}$$

Q and R are system noise and observation noise covariance matrix[8,9], E[*] is expectation operator, if k=j, δ_{kj}=1; else δ_{kj}=0.

Innovation was the difference between observation data and forecast data, formula :

$$d(k)=Y(k)-H(k)\hat{X}(k|k-1) \tag{3}$$

In reference [4], the author proved that linear optimal filter innovation sequence($\{d(k)\}$)which based on Kalman filter and prediction was zero mean Gauss white noise.

Multi-step innovation($\{d_{\mathrm{T}}(k)\}$) was the difference between observation variables Y(k) and Multi-step forecast variables $\hat{\mathrm{X}}(k/k-1)$ at T step.

$$d_T(k)=Y(k)-H(k)\hat{X}(k|k-T) \tag{4}$$

Here, T was an integer multiple of the system operating cycle. If the target was not maneuvering, for non maneuvering target model, $\{d_{\mathrm{T}}(k)\}$ was also the zero mean Gauss white noise characteristic, the maneuver detection method described in this paper was also based on the characteristic.

3. Maneuver Detection Method

Detection of maneuvering targets was a decision problem, which could be formulated as a hypothesis testing problem.

H0:non maneuvering, H1:maneuvering

According to the characteristics of observation residuals and control inputs, In maneuvering target tracking, we could use testing using the χ^2 detection methods and or quasi χ^2 detection methods . Under the condition of H0 (detection maneuver startment), we estimated whether or not the Statistic was χ^2 distribution. Assuming that under H0 conditions, ε was χ^2 distribution $\varepsilon\sim\chi_n^2$ with degree of freedom is n, and the next equation was is set up, it indicated that the target is maneuvering.

$$\varepsilon>\lambda=\chi_n^2(\alpha) \tag{5}$$

Here $1-\alpha$ was confidence value.

Normalized residual squared was $\varepsilon_k=d'(k)S_k^{-1}d(k)$, $S_k=\mathrm{cov}(d(k))$. In order to reduce the probability of false alarm and false alarm, this used the method of attenuation memory statistics to test the hypothesis. The sum of memory decay was ε_k^{ρ}.

$$\varepsilon_k^{\rho}=\sum_{j=1}^{k}\rho^{k-j}\varepsilon_j=\rho\varepsilon_{k-1}^{\rho}+\varepsilon_k,\ 0<\rho<1,\ s_{\rho}=\frac{1}{1-\rho} \tag{6}$$

Here, ρ was attenuation coefficient, s_ρ was the length of the sum of Attenuation memory equivalent window. Because ε_k was χ^2 distribution $\varepsilon \sim \chi_n^2$ with degree of freedom is n($\varepsilon_k \sim \chi_n^2$, n = dim(d)). ε_k^ρ was t not χ^2 distribution, but it could be approximated as a χ^2 variable by weight matching. That was approximate[10]:

$$\varepsilon_k^\rho \sim \frac{1}{1+\rho}\chi_{n_\rho}^2, \quad n_\rho = n\frac{1+\rho}{1-\rho} \tag{7}$$

In application, we used n = 3 when tracking in spherical coordinate system. At this point, the maneuvering start can be detected according to the equation (6).

4. Maneuver Detection Delay Time

Generally, the problem of hypothesis testing was expected to get as many measurements as possible to improve the detection probability. But for the problem of maneuver detection, we hoped to get the judgment as soon as possible. That was the minimum delay. In this way, we faced two conflicting goals. That was to increase the detection probability and tried to delay the decision, at the same time we tried to reduce the delay to avoid tracking lag. So, it was very important to obtain the detection delay time accurately under the constant false alarm rate.

According to equation (6), It could be seen that the calculated value ε_k^ρ was related to the amplitude of maneuver (the value of innovation) and the length of the window（s_ρ）. Maneuver detection delay time which was χ^2 distribution was the function of the system model, maneuvering range, threshold value（α） and window length(equivalent to the size of the attenuation coefficient). Maneuver detection delay time which based on multi-step innovation was the function of the model of the system, the range of motion (with the step length), the confidence and the length of the window.

Confidence value was given according to system requirements, If the α is small, confidence value is high, the threshold value of maneuvering target is higher, therefore, the detection delay is large. So, α was inversely proportional to the detection delay. Maneuvering amplitude also affected detection delay, the large maneuver made the value of the innovation sequence larger, So the statistical value ε_k^ρ has a larger value. For a fixed detection threshold, it was easy to detect maneuver earlier. In fact, Since the target was unknown, thus, the amplitude of maneuver was unknown in advance, it could not be used for system optimization. When the window length s_ρ was large, randomness of $\varepsilon_k^\rho / s_\rho$ was very weak, detection lag was larger. At the same time, if the step of multi-step innovation was larger, it would also bring detection lag, so the selection of the step size of multi-step innovation, and the window length (attenuation coefficient) could be selected to optimize the system.

5. Simulation Analysis

Considering the complexity of the optimal problem, the simulation method was used here, to guide the choice of step length and attenuation coefficient.

The target simulation route was from the uniform straight line to a certain fixed load to do uniform acceleration motion, a total of 5 target route simulation were used. Target overload capacity was from 1.0g to 5.0g, for each route, the attenuation coefficient ranged from 0.1 to 0.8. Under different overload conditions, the relationship between the minimum detection delay time and attenuation coefficient was shown in Figure 1. It could be seen that the smaller the attenuation coefficient, the longer the delay time, This was also consistent with theoretical analysis.

For each of the above routes in fixed attenuation coefficient, the step size of the multi-step innovation was set from 10 to 60 (the number of operating cycles),and simulated. Simulation results showed: for a highly maneuverable target course (as described in this article), there was no change in the detection delay time under different step sizes, But it could significantly reduce the false alarm rate of target maneuvering detection, as shown in figure 2.

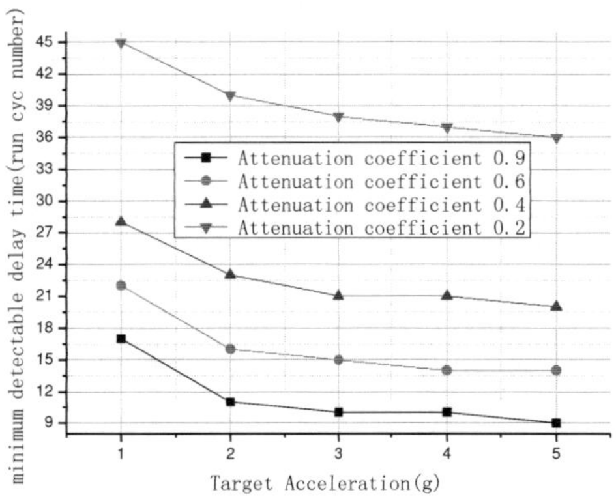

Figure 1. Relationship between minimum detection delay time and attenuation coefficient

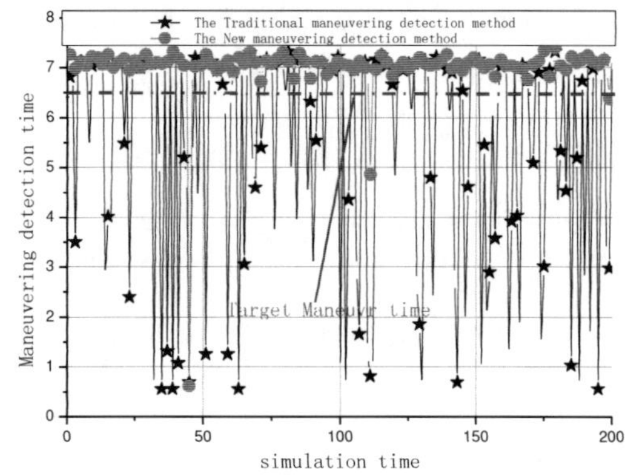

Figure 2. Comparison of multi-step innovation detection method and traditional maneuver detection method

For weak maneuvering routes, the same transition from uniform straight line to a fixed overload to do uniform acceleration movement, but the target overload is only 0.2g. In the fixed attenuation coefficient conditions, the minimum detection delay time and the relationship between the steps shown in Figure 3. It can be seen that there is an optimal prediction step for the fixed attenuation coefficient.

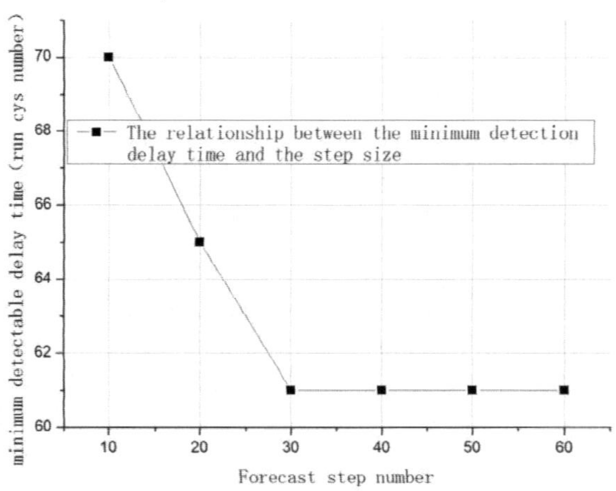

Figure 3. Relationship between the minimum detection delay time and the multi-step forecasting step size

6. Conclusion

Anti-ship missiles an important direction of development is faster and faster, leaving the end of the defense system response time is getting shorter and shorter. An effective way to improve the handling capacity of anti-ship missiles is to improve the utilization rate of the tracking system for tracking data, that is, to improve the data rate of the end defense system. But the improvement of the data rate will reduce the traditional signal-to-noise ratio based on the single-step new machine

IOP Conf. Series: Materials Science and Engineering **361** (2018) 012026 doi:10.1088/1757-899X/361/1/012026

detection window. Based on this, this paper presents a maneuver detection method based on multi-step new information, and adopts the maneuver detection method based on attenuation memory. According to the given maneuver detection threshold, the analysis and simulation show that the false alarm rate can be significantly reduced by the multi-step dynamic detection method. For the simulation route, the minimum detection delay time and attenuation coefficient are given, and the relationship between the delay time and the prediction step is given.

References

[1] Guo-liang Xu. Maneuvering target tracking algorithm [J]. *Journal of intelligence command and control system and simulation technology.* 2002 8 42-56.

[2] Zhangsong Shi and Jun Xie. Maneuvering detection algorithm characteristic simulation [J]. *Computer Simulation.* 2007 24 90-94.

[3] Chao Gao and Xiaobo Deng. High speed maneuvering target detection algorithm research [J]. *Computer simulation.* 2014 31 1-5.

[4] Hongli Yao, Xiao Gao and Keyu Tian. Based on the attenuation memory n new rate target maneuver detection and flying spot [J]. *Journal of intelligence command and control system and simulation technology.* 2004 26 25-29.

[5] Liping Song, Hongbing Ji and Xinbo Gao. Based on higher-order cumulant, a new method of target maneuver detection [J]. *Journal of electronics.* 2004 32 154-6.

[6] Hai Li, Renbiao Wui and Xiaohan Wang. Aerial maneuvering target detection based on nonlinear least squares method [J]. *Journal of electronics &information technology.* 2012 9 2143-47.

[7] HaiYan Shang, Yabin Zhang, Hongtao Su, Shouhong Zhang and Peng lang Shui. Long observation time maneuvering target detection method [J]. *Journal of systems engineering and electronics.* 2006 28 1779-82.

[8] Rui Zhang and Linan Shi. Research of maneuvering target tracking based on extended kalman filter [J]. *Aerospace control.* 2012 30 12-17.

[9] Shuai Yue, Lingjiang Kong, Jianyu Yang and Wei Yi. Kalman dynamic programming method of tracking maneuvering target detection [J]. *Journal of modern radar.* 2011 6 58-63.

[10] Guoliang Xu and Yajuan Deng. Maneuvering target modeling and maneuver detection algorithm [J]. *Journal of intelligence command and control system and simulation technology.* 2005 27 81-83.